Real Math
PRACTICE WORKBOOK
Grade 4

Stephen S. Willoughby

•

Carl Bereiter

•

Peter Hilton

•

Joseph H. Rubinstein

•

Joan Moss

•

Jean Pedersen

Columbus, OH

The **McGraw·Hill** Companies

SRAonline.com

Send all inquiries to:
SRA/McGraw-Hill
4400 Easton Commons
Columbus, OH 43219

ISBN 0-07-603738-X

4 5 6 7 8 9 BCH 12 11 10 09 08 07

The McGraw·Hill Companies

Practice Lessons

Practice Lessons

Name _____ **Date** _____

Estimating

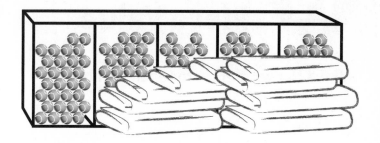

A rack in the gym at David's school holds tennis balls. It is partially covered so that the exact number of balls in the rack cannot be counted.

1 Write your best estimate of how many tennis balls there are altogether.

_____

2 How did you determine your estimate?

_____

_____

_____

3 David's friends helped him carry away some balls. There were 5 children who took 5 balls each. How many balls did they take altogether?

_____

4 Make an estimate of how many balls there are now.

_____

5 Then 5 more children came. They each took 5 balls. So far, 10 children have taken 5 balls. How many balls is that?

_____

6 Make an estimate of how many balls are left.

_____

7 More children came and took balls. So far, 15 children have taken 5 balls each. How many balls is that?

_____

8 Make an estimate of how many balls are left.

_____

Name _____ **Date** _____

Place Value

Write the word form for each number.

1 2,562 _____

2 35,192 _____

3 12,014 _____

4 104,685 _____

Write the numbers in standard form.

5 300 + 40 + 5 _____

6 1 + 60 + 900 + 7,000 _____

7 100 + 3,000 + 60,000 _____

8 7,000,000 + 50,000 _____

Copy and complete the table. Use the digits in the first column to find the greatest and the least numbers for each combination of digits.

	Use These Digits	Greatest Number	Least Number
9	8, 9		
10	6, 3, 7, 8		
11	5, 5, 4, 4, 8, 1		

Arrange the numbers below in a row to make six different numbers, in order from greatest to least.

12 2, 4, and 9

_____

13 1, 5, and 7

_____

Real Math • Grade 4 • *Practice*

Name _____ **Date** _____

Numerical Sequence

Count on. Write the missing numbers.

1 23, 24, 25, _____, _____, _____, _____, _____, _____, 32

2 7,085, 7,086, 7,087, _____, _____, _____, _____, 7,093

3 99,996; 99,997; _____; _____; _____; _____; 100,002

Count back. Write the missing numbers.

4 672, 671, 670, _____, _____, _____, _____, _____, 664

5 2,908; 2,907; 2,906; _____; _____; _____; _____; _____;

_____; _____; 2,898

6 104,398; 104,397; 104,396; _____; _____; _____; _____;

_____; _____; 104,389

Count on or back. Write the missing numbers.

7 45,996; 45,997; _____; _____; _____; _____; _____; 46,003

8 Look at the sequence.

1045, 1044, 1043, 1042, _____, _____, _____

Would you count on or back? Explain.

_____

_____

9 Write three numbers that come before and three
numbers that come after these numbers.

_____; _____; _____; 12,490; 12,491; 12,492; _____; _____;

_____

10 Explain how you know to count on or back.

_____

Rounding

Round each number to the nearest ten.

1 56 _____

2 4,970 _____

3 455 _____

4 3,991 _____

Round each number to the nearest hundred.

5 2,459 _____

6 394 _____

7 33 _____

8 3,991 _____

Round each number to the nearest thousand.

9 7,552 _____

10 993 _____

11 36,220 _____

12 247,848 _____

13 6,386,299 _____

14 37,716 _____

Solve these word problems by rounding.

15 Malik wants to buy a baseball glove that costs $35 and a ball that costs $18. He has $50. Does he have enough money to buy the glove and the ball?

_____

16 Lee's new truck can tow 5,600 pounds of cargo. His boat weighs 3,895 pounds. His trailer weighs 1,050 pounds. Can Lee tow both items at the same time?

_____

17 Explain how you used rounding to solve Problem 16.

_____

Real Math • Grade 4 • *Practice*

Name _____ Date _____

Practicing Addition and Subtraction

Add to solve for *n*.

1 $6 + 7 = n$ _____

2 $11 + 5 = n$ _____

3 $n = 5 + 2 + 3$ _____

4 $n = 3 + 2 + 5$ _____

Add.

5
$$\begin{array}{r} 9 \\ + 8 \\ \hline \end{array}$$

6
$$\begin{array}{r} 7 \\ + 6 \\ \hline \end{array}$$

7
$$\begin{array}{r} 6 \\ + 8 \\ \hline \end{array}$$

8
$$\begin{array}{r} 2 \\ 4 \\ + 5 \\ \hline \end{array}$$

9
$$\begin{array}{r} 9 \\ 2 \\ + 5 \\ \hline \end{array}$$

Subtract to solve for *n*.

10 $10 - 3 = n$ _____

11 $10 - 7 = n$ _____

12 $n = 16 - 16$ _____

13 $n = 15 - 5$ _____

14 $17 - 6 = n$ _____

15 $n = 20 - 7$ _____

Subtract.

16
$$\begin{array}{r} 14 \\ - 8 \\ \hline \end{array}$$

17
$$\begin{array}{r} 15 \\ - 7 \\ \hline \end{array}$$

18
$$\begin{array}{r} 12 \\ - 4 \\ \hline \end{array}$$

19
$$\begin{array}{r} 8 \\ - 7 \\ \hline \end{array}$$

20
$$\begin{array}{r} 13 \\ - 6 \\ \hline \end{array}$$

Solve.

21 Ashley has 8 feet of yellow yarn and 6 feet of green yarn.

    a. How much yarn does she have altogether?

_____

    b. How much more yellow yarn does she have than green yarn?

_____

Name _____ Date _____

Function Machines

Complete each table. Pay attention to what each function rule means.

①

−3	
In	Out
3	0
6	3
9	6
12	
15	
18	

②

+4	
In	Out
0	4
1	
2	
3	
4	
5	

③

+8	
In	Out
	9
2	
4	
	13
	15
8	

④

Rule: _____	
In	Out
10	2
12	4
14	6
16	
18	

Use the information below to help you answer these questions.

| $6.00 | $3.00 | $15.00 | $4.00 |

Sam buys a teddy bear, and Gabriella buys a hat and a book.

⑤ Who spent more money? Explain.

_____

⑥ How much more money did Sam spend than Gabriella?

_____

⑦ Gabriella started with $12. Can she also buy a brush?
Why or why not?

_____

Real Math • Grade 4 • *Practice*

Missing Addends

Complete the table.

	Ryan has	The sum is	Tara has		Drew has	The sum is	Emma has
①	4	11		**②**	7	16	

Solve for n.

③ $n + 5 = 12$ _____

④ $12 - n = 6$ _____

⑤ $3 + n = 8$ _____

⑥ $n - 4 = 11$ _____

⑦ $n + 8 = 14$ _____

⑧ $18 - n = 9$ _____

⑨ $8 + n = 17$ _____

⑩ $n - 2 = 10$ _____

⑪ $n + 2 = 18$ _____

⑫ Matthew has 16 math problems for homework. He has 9 left to solve. How many problems has he already solved? $16 - n = 9$

_____

⑬ A shelter had 8 beagle puppies. So far, 6 of the puppies have been adopted. How many puppies are left?
$8 - 6 = n$

_____

Solve.

⑭ John bought a plant. It grew 3 inches. It is now 9 inches tall. How tall was the plant when John bought it?

_____

⑮ Diego has $24. He spent some money at the store. How much does he have now?

_____

Name _____ **Date** _____

Perimeter

Find the perimeter.

1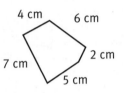

4 cm

3 cm

_____

2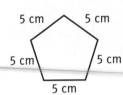

5 cm   5 cm

5 cm   5 cm

5 cm

_____

3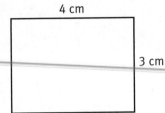

4 cm   6 cm

7 cm   2 cm

5 cm

_____

4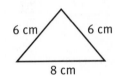

6 cm   6 cm

8 cm

_____

Solve. Drawing diagrams might help.

5 Jasmine is making a rectangular-shaped garden. One side of the garden is 8 feet. The perimeter of the garden is 28 feet. What is the length of the other side of the garden? How did you find the length of the side?

_____

_____

6 In an equilateral triangle, all three sides are equal. One side of an equilateral-shaped flag is 7 feet. What is the perimeter of the flag?

_____

7 How many squares can you draw that have a perimeter of 16 feet? Draw them all on graph paper using whole numbers for the lengths of the sides.

_____

Real Math • Grade 4 • *Practice*

Name _____ Date _____

Using Maps and Charts

Estimate the driving distances, in miles, between the following United States cities. For each pair of cities four distances are given in miles, but only one is correct. Select the correct distance.

❶ Memphis to Jackson

    a. 2,240   **b.** 1,950   **c.** 210   **d.** 100

❷ Orlando to Washington, D.C.

    a. 2,300   **b.** 250   **c.** 1,400   **d.** 850

❸ New Orleans to Orlando

    a. 1,540   **b.** 650   **c.** 1,130   **d.** 250

❹ Jackson to Little Rock

    a. 260   **b.** 2,450   **c.** 1,800   **d.** 650

Study the table. Then answer the questions for the zoos listed in the table.

Major Public Zoos in the United States			
Zoo	Yearly Attendance	Acres	Number of Animal Species
Brookfield	2,000,000	215	400
Columbus	1,300,000	90	700
St. Louis	2,900,000	83	800
San Diego	3,300,000	100	800

❺ Which zoo has the greatest attendance?

_____

❻ Does the zoo in Problem 5 also have the greatest number of acres?

_____

❼ Which two zoos have about the same number of acres?

_____

Name _____ Date _____

Multidigit Addition

Add.

① 65
  + 23

② 45
  + 45

③ 90
  + 90

④ 66
  + 66

⑤ 66
  + 65

⑥ 48
  + 92

⑦ 200
  + 200

⑧ 200
  + 199

⑨ 199
  + 199

⑩ 805
  + 45

⑪ 387
  + 795

⑫ 74,315
  + 26,949

⑬ 2,893,417
  + 6,484,385

⑭ 27
  36
  + 45

⑮ 216
  387
  + 691

Circle each correct answer.

⑯ 61 + 61 =
  a. 610
  b. 6,161
  c. 122

⑰ 57 + 182 =
  a. 652
  b. 239
  c. 5,720

⑱ 198 + 213 =
  a. 411
  b. 3,110
  c. 185

⑲ 7,835 + 4,935 =
  a. 117,765
  b. 3,100
  c. 12,770

Solve the following problems.

⑳ Kenji's school collected books for an after-school program. The third-grade collected 146 books, the fourth-grade collected 219 books, and the fifth-grade collected 184 books. How many books were collected altogether?

_____

㉑ A library in Kenji's neighborhood agreed to donate 125 books. Was the total number of books donated more than 1,000? Explain. (Hint: Use your answer from Problem 20.)

_____

Real Math • Grade 4 • *Practice*

Name _____ **Date** _____

Multidigit Subtraction

Subtract. Use shortcuts when you can.

1   74   **2**   88   **3**   92   **4**   98   **5**   44
    − 23   − 80   − 37   − 69   − 25

6   683   **7**   784   **8**   2,976   **9**   1,783   **10**   436,927,520
    − 421   − 652   − 1,849   − 599   − 295,147,816

Add or subtract. Use shortcuts when you can.

11   304   **12**   708   **13**   5,923   **14**   304   **15**   4,987
    − 178   − 224   + 2,077   + 267   − 3,999

Circle the correct answer.

16   692 − 45 =   **17**   595 − 340 =   **18**   958 − 692 =   **19**   716 + 638 =

   a. 657   **a.** 144   **a.** 346   **a.** 1,354

   b. 653   **b.** 235   **b.** 266   **b.** 1,344

   c. 647   **c.** 255   **c.** 250   **c.** 1,454

Answer the following questions.

20 Andrew's soccer team is having three fundraisers so they can buy new uniforms. The new uniforms cost $200. At the first fundraiser the team made $59, and at the second fundraiser they made $77. How much does the team need to make at the third fundraiser?

_____

21 The heaviest recorded great white shark weighed 7,731 pounds. The heaviest recorded tiger shark weighed 2,043 pounds. How much more did the great white shark weigh than the tiger shark?

_____

Name _____ Date _____

Multidigit Addition and Subtraction

Add or subtract. Use shortcuts when you can.

① 42
   + 47

② 77
   − 24

③ 50
   − 15

④ 35
   + 15

⑤ 425
   − 72

⑥ 5,500
   − 3,400

⑦ 5,500
   − 3,401

⑧ 5,501
   − 3,402

⑨ 5,457
   + 3,002

⑩ 476,520
   + 534,936

Complete the table.

As part of a year-long school fundraiser, Hannah's and Daniel's classes raised money. They kept a table of the amount of money they raised.

	Hannah's Class	**Daniel's Class**
First Half of the Year	1,286	814
Second Half of the Year	1,498	**⑫** _____
Total Raised	**⑪** _____	1,766

Answer the following questions based on the table.

⑬ Which class raised more money? How much more did they raise?

_____

⑭ How much more money did Hannah's class raise than Daniel's class in the first half of the year? _____

⑮ How much was raised by both classes altogether?

_____

⑯ Hannah's class set a goal of $2,000. Did the class reach their goal? What was the difference between their goal and the actual amount raised?

_____

Name _____ Date _____

Using Relation Signs

Write <, >, or = to make each statement true.

1 84 _____ 88

2 6 + 7 _____ 23

3 200 _____ 127 + 79

4 38 _____ 11 + 27

5 66 _____ 70 − 6

6 70 − 1 _____ 68

7 16 + 16 _____ 116

8 4 + 3 _____ 2 + 3

9 16 + 7 _____ 7 + 16

10 16 − 7 _____ 16 + 7

11 45 _____ 25 + 20

12 20 + 20 _____ 50 − 5

Solve. Use the table to determine your answers.

Kia's House		
Room	**Length**	**Width**
Bathroom	8 feet	4 feet
Kitchen	13 feet	14 feet
Dining Room	20 feet	10 feet
Living Room	23 feet	27 feet

13 Kia has 150 feet of wallpaper border. She has
4 rectangular rooms where she wants to hang the
border. Using the table above, does she have enough
border to cover the perimeter of

a. the bathroom and the kitchen? _____

b. the kitchen and the dining room? _____

c. the dining room and the living room? _____

Name _____ Date _____

Addition and Subtraction with Hidden Signs

Circle the correct answer. Each represents a missing number. One answer is correct in each problem.

❶   76
+ 2◯

   a. 90

   b. 100

   c. 110

❷   302
+ ◯9

   a. 381

   b. 481

   c. 301

❸   626
+ 5◯0

   a. 126

   b. 1,056

   c. 1,146

❹   275
− ◯4

   a. 91

   b. 191

   c. 291

❺   7◯
− 56

   a. 9

   b. 19

   c. 29

❻   83◯
+ 79

   a. 817

   b. 917

   c. 1,009

❼   2◯
+ 424

   a. 600

   b. 700

   c. 800

❽   746
− 40◯

   a. 347

   b. 237

   c. 337

❾   7◯0
+ 125

   a. 845

   b. 945

   c. 765

❿   98
− 3◯

   a. 64

   b. 72

   c. 94

⓫ 2◯8
− 809

   a. 2,869

   b. 869

   c. 1,869

⓬ 4760
− 892

   a. ◯868

   b. ◯,968

   c. ◯878

Solve.

⓭ What number makes the sum equal 100?

       71
    + 2◯

⓮ Fill in the ◯ to make two addition problems.

  Problem 1   Problem 2

    3◯   3◯
  + 2◯   + 2◯
    4◯   ◯3

Name _____ Date _____

Approximation Applications

Select the best approximation for each problem, and circle it. You do not have to do the calculations.

1 Jamestown, Virginia, was established in 1607. About how many years ago was that?

   a. 400   **b.** 40   **c.** 140

2 France recognized America's independence in 1778. About how many years ago was that?

   a. 25   **b.** 225   **c.** 525

3 The United States of America negotiated the Louisiana Purchase from France in 1803. About how many years ago was that?

   a. 100   **b.** 10   **c.** 200

4 X-rays were discovered by German physicist Wilhelm Roentgen in 1895. About how many years ago was that?

   a. 100   **b.** 10   **c.** 50

5 In 2000 the population of Kansas City, Missouri, was 441,545, and the population of Memphis, Tennessee, was 650,100. About how many more people lived in Memphis than in Kansas City?

   a. 200,000   **b.** 2,000,000   **c.** 20,000

6 In 2000 the population of Dallas, Texas, was 1,188,580, and the population of Indianapolis, Indiana, was 781,870. About how many more people lived in Dallas than in Indianapolis?

   a. 4,000   **b.** 40,000   **c.** 400,000

7 The air distance between Chicago, Illinois, and Salt Lake City, Utah, is 1,260 miles. The air distance between Chicago and San Francisco, California, is 1,858 miles. Chicago is about how much farther from San Francisco than it is from Salt Lake City?

   a. 60 miles   **b.** 600 miles   **c.** 6,000 miles

Name _____ Date _____

Making Inferences

Turtle Weights	
Turtle	**Weight (pounds)**
Atlantic leatherback	1,018
Green sea	783
Loggerhead	568

Use the table above to choose the best answer to the following questions. Circle your answer.

❶ About how heavy is an Atlantic leatherback turtle?

    a. 500 pounds   **b.** 800 pounds   **c.** 1,500 pounds

❷ About how much more does a Green sea turtle weigh than a Loggerhead turtle?

    a. 300 pounds   **b.** 100 pounds   **c.** 200 pounds

Isabella asked her classmates to toss a 1–6 *Number Cube*. She listed the times the cube landed on each number, but only half of what she listed is shown in the table.

Number	
1	6
2	8
3	14

Use the information from the table to give exact answers to the following questions.

❸ How many students rolled a 3? _____

❹ What number did 8 students roll? _____

Use the same table to make good estimates to the following questions.

❺ About how many students rolled a 6?

_____

❻ About how many students rolled a 4, 5, or 6?

_____

Integers

Read the information below. Then answer each question.

The students in Miss Orta's reading group have set a goal to read five books each per month.

If a student reads fewer than five books in one month, the number of books not read is recorded with a negative number. If a student reads more than five books in one month, the number of extra books read is recorded with a positive number. Miss Orta's students kept a table for ten months.

Say whether each student read more or less than the goal over the ten months.

1	Tomas	+1, 0, −1, +2, 0, −1, +3, 0, 0, +1	
2	Jessica	−2, +1, 0, −1, +1, +2, −2, 0, +1, +1	
3	Mia	−1, −1, +1, 0, +1, −2, 0, −2, −1, −1	
4	Noah	+4, +2, 0, +1, −1, −1, 0, −2, +1, 0	
5	Jacob	+1, +1, 0, −2, −1, 0, −1, −1, +1, −2	

Answer the following questions.

6 List the students who have read more than the goal so far.

_____

7 List the students who have read less than the goal so far.

_____

8 How many books will Jacob need to read next month to be exactly at his goal?

_____

9 Does Jessica need to read any more books to reach her goal? How many fewer books would Jessica have needed to read in the last ten months to be exactly at her goal?

_____

Name _____ Date _____

Adding and Subtracting Integers

Use what you have learned to solve the following exercises.

1 6 + 5 = _____   **2** 6 − 5 = _____   **3** −6 + 5 = _____   **4** −6 − 5 = _____

5 7 − 7 = _____   **6** 7 − (−7) = ___   **7** 7 + 7 = _____   **8** 7 + (−7) = ___

Circle each correct answer.

9 −10 + 4

   a. 6

   b. −6

   c. 14

10 6 + (−9)

   a. −3

   b. −15

   c. 3

11 −7 − (−3)

   a. −4

   b. 4

   c. −10

12 −2 + (−6)

   a. 4

   b. 8

   c. −8

Solve the following problems.

13 The temperature at 8:00 P.M. is 13 degrees. The temperature falls 10 degrees between 8:00 P.M. and 12:00 A.M. What is the temperature at 12:00 A.M.?

_____

14 The temperature falls another 6 degrees between 12:00 A.M. and 7:00 A.M. What is the temperature at 7:00 A.M.? (Hint: Use Problem 13.)

_____

15 A fish swims in the ocean 9 feet below sea level. The fish swims down 5 feet. How far below sea level is the fish now?

_____

16 When Kayla awoke the temperature was −8 degrees. The temperature has fallen 15 degrees since last night at 10:00 P.M. What was the temperature at 10:00 P.M. last night?

_____

Name _____ Date _____

Understanding Multiplication

Solve for *n*. Use any method.

1 5 × 4 = n _____   **2** 3 × 2 = n _____   **3** 4 × 4 = n _____

4 6 × 3 = n _____   **5** 2 × 9 = n _____   **6** 5 × 2 = n _____

7 7 × 8 = n _____   **8** 6 × 4 = n _____   **9** 9 × 5 = n _____

10 2 × 4 = n _____   **11** 7 × 1 = n _____   **12** 8 × 3 = n _____

13 7 × 7 = n _____   **14** 4 × 0 = n _____   **15** 3 × 9 = n _____

16 1 × 8 = n _____   **17** 7 × 2 = n _____   **18** 10 × 6 = n _____

19 11 × 3 = n _____   **20** 3 × 5 = n _____   **21** 12 × 4 = n _____

Solve for *n*. Write *yes* if the two equations show the commutative property.

22 3 × 8 = n   **23** 5 × 6 = n   **24** 2 × 3 = n
  6 × 4 = n   6 × 5 = n   6 × 1 = n

_____   _____   _____

25 4 × 3 = n   **26** 9 × 2 = n   **27** 6 × 6 = n
  3 × 4 = n   2 × 9 = n   9 × 4 = n

_____   _____   _____

Solve.

28 Carmen needs to find a CD case that will hold 30 of her CDs.

a. She found one case that has 8 pages. Each page holds 4 CDs. Will the case hold 30 CDs? Explain.

_____

b. If the case has 4 pages and each page holds 8 CDs, will the case hold 30 CDs? Explain.

_____

Name _____ **Date** _____

Multiplying by 0, 1, 2, and 10

Multiply to solve for *n*.

1 $0 \times 4 = n$ _____

2 $5 \times 10 = n$ _____

3 $5 \times 2 = n$ _____

4 $2 \times 7 = n$ _____

5 $10 \times 3 = n$ _____

6 $7 \times 1 = n$ _____

7 $2 \times 9 = n$ _____

8 $6 \times 2 = n$ _____

9 $1 \times 8 = n$ _____

10 $6 \times 0 = n$ _____

11 $8 \times 2 = n$ _____

12 $1 \times 10 = n$ _____

13 $3 \times 2 = n$ _____

14 $5 \times 1 = n$ _____

15 $12 \times 2 = n$ _____

Find each product.

16
$$\begin{array}{r} 4 \\ \times\,2 \\ \hline \end{array}$$

17
$$\begin{array}{r} 9 \\ \times\,1 \\ \hline \end{array}$$

18
$$\begin{array}{r} 10 \\ \times\,0 \\ \hline \end{array}$$

19
$$\begin{array}{r} 2 \\ \times\,2 \\ \hline \end{array}$$

20
$$\begin{array}{r} 3 \\ \times\,1 \\ \hline \end{array}$$

21
$$\begin{array}{r} 10 \\ \times\,6 \\ \hline \end{array}$$

22
$$\begin{array}{r} 9 \\ \times\,10 \\ \hline \end{array}$$

23
$$\begin{array}{r} 0 \\ \times\,7 \\ \hline \end{array}$$

24
$$\begin{array}{r} 9 \\ \times\,2 \\ \hline \end{array}$$

25
$$\begin{array}{r} 9 \\ \times\,6 \\ \hline \end{array}$$

26
$$\begin{array}{r} 5 \\ \times\,6 \\ \hline \end{array}$$

27
$$\begin{array}{r} 1 \\ \times\,5 \\ \hline \end{array}$$

28
$$\begin{array}{r} 5 \\ \times\,4 \\ \hline \end{array}$$

29
$$\begin{array}{r} 10 \\ \times\,2 \\ \hline \end{array}$$

30
$$\begin{array}{r} 9 \\ \times\,0 \\ \hline \end{array}$$

Solve each problem.

31 Raul has a dog and a cat. The weight of his dog is 10 times the weight of his cat. His cat weighs 8 pounds. How much does his dog weigh?

_____

32 Is it possible to multiply a whole number by 2 and get 15? Explain how you know.

_____

Name _____ **Date** _____

Multiplying by 5 and 9

Find each product.

1 $8 \times 9 = n$ _____ **2** $7 \times 9 = n$ _____ **3** $5 \times 6 = n$ _____ **4** $7 \times 5 = n$ _____

5 $9 \times 2 = n$ _____ **6** $2 \times 5 = n$ _____ **7** $10 \times 3 = n$ _____ **8** $5 \times 5 = n$ _____

9 $9 \times 7 = n$ _____ **10** $9 \times 4 = n$ _____ **11** $10 \times 4 = n$ _____ **12** $5 \times 7 = n$ _____

Solve each problem.

13 Tickets for the science museum cost $5 for children and $9 for adults. Mr. Smith bought 3 child tickets and 2 adult tickets. How much did Mr. Smith spend on tickets?

_____

14 At the science museum, postcards cost $1 each. For every four postcards you buy, you get one free. Alexis picked out 10 postcards. How much did she spend?

_____

15 Can you multiply a whole number by 5 and get 56? Explain.

_____

16 Multiply each number in the chart below by 9. Then add the digits in each product.

Complete the following table.

	1	2	3	4	5	6	7	8	9	10
× 9										
Sum of Digits										

a. What pattern do you see in the table above?

_____

b. Is it possible to multiply a whole number by 9 and get 109? Explain how you know.

_____

Name _____ **Date** _____

Square Facts

Find the area. Remember that to find the area of a square, you multiply the length of a side by itself.

6 cm

Square A

1 What is the area of Square A? _____

2 What is the area of Square B? _____

8 cm

3 What is the area of Square C? _____

Square B

4 What is the area of Square A plus the area of Square B?

_____

10 cm

5 Describe the relationship among the areas of the three squares.

_____

Square C

Complete the tables below.

	Length of Side	Area of Square
6	10 in.	
7	9 in.	
8	8 in.	
9	7 in.	
10	6 in.	

	Length of Side	Area of Square
11	5 in.	
12	4 in.	
13	3 in.	
14	2 in.	
15	1 in.	

Solve.

16 Mrs. Palermo bought 10 boxes of tiles to cover the floor in her sunroom. Each box holds 8 one-square foot tiles. The sunroom is a square with a width of 9 feet. Did Mrs. Palermo buy enough tiles? Explain.

_____

Name _____ **Date** _____

Multiplying by 3, 4, 6, and 8

Multiply to solve for *n*.

1 $2 \times 3 = n$ _____   **2** $5 \times 6 = n$ _____   **3** $8 \times 5 = n$ _____

4 $4 \times 5 = n$ _____   **5** $2 \times 8 = n$ _____   **6** $0 \times 7 = n$ _____

7 $6 \times 8 = n$ _____   **8** $3 \times 8 = n$ _____   **9** $5 \times 1 = n$ _____

10 $7 \times 1 = n$ _____   **11** $9 \times 2 = n$ _____   **12** $3 \times 7 = n$ _____

13 $0 \times 10 = n$ _____   **14** $7 \times 6 = n$ _____   **15** $8 \times 8 = n$ _____

16 $5 \times 7 = n$ _____   **17** $6 \times 9 = n$ _____   **18** $12 \times 3 = n$ _____

19 $8 \times 4 = n$ _____   **20** $4 \times 4 = n$ _____   **21** $5 \times 5 = n$ _____

22 $7 \times 8 = n$ _____   **23** $9 \times 6 = n$ _____   **24** $8 \times 7 = n$ _____

25 $5 \times 9 = n$ _____   **26** $3 \times 9 = n$ _____   **27** $6 \times 2 = n$ _____

Solve each problem.

28 Aubrey and Sarah play on the same basketball team. In the last game Sarah scored 4 times as many points as Aubrey. Is it possible that Sarah scored 17 points? Explain.

_____

_____

29 Darian organized his photos in an album. He filled 9 pages with photos. Darian placed 4 photos on each side of a page. How many photos are in Darian's album?

_____

30 William multiplied two whole numbers and got 40 as the answer. His brother multiplied two different whole numbers and got the same answer. Is this possible? Explain.

_____

_____

Name _____ **Date** _____

Multiplication and Addition Laws

Solve the following exercises. Use shortcuts whenever possible.

1 $528 \times 0 =$ _____

2 $528 + 0 =$ _____

3 $528 \times 1 =$ _____

4 $4 \times 68 =$ _____

5 $90 \times 5 =$ _____

6 $93 \times 5 =$ _____

7 $43 + 43 =$ _____

8 $43 \times 2 =$ _____

9 $68 \times 4 =$ _____

10 $86 + 86 =$ _____

11 $43 \times 4 =$ _____

12 $98 \times 1 =$ _____

13 $12 \times 9 \times 6 \times 3 \times 1 \times 0 =$ _____

14 $(5 \times 4) \times (3 \times 2) \times 1 =$ _____

15 $0 \times 15 \times 20 \times 5 \times 45 =$ _____

16 $1 \times 2 \times 3 \times 4 \times 5 =$ _____

17 $(26 + 34) \times 3 =$ _____

18 $(26 \times 3) + (34 \times 3) =$ _____

19 $1 + 2 + 3 + 4 + 5 + 5 + 6 + 7 + 8 + 9 =$ _____

Solve each problem.

20 At the video rental store, there are 40 shelves of movies. There are 68 movies on each shelf. How many movies are there altogether? (Hint: Does the answer to Problem 4 help?)

_____

21 An employee at the video rental store counted the number of movies on each shelf. There were 10 shelves that had only 65 movies each. Of the remaining shelves, 15 shelves had 70 movies, and 15 shelves had exactly 68 movies. How many movies are really in the store?

_____

22 Brandon has 3 movies he wants to rent. It costs $3.50 to rent 1 movie. Help Brandon figure out how much he will have to pay to rent the 3 movies.

_____

23 Explain how to use the distributive property to help you solve 217×4.

_____

Name _____ **Date** _____

Multiplying by 11 and 12

Answer the following questions.

1 Ms. Porter is planning a birthday party for her niece. She baked 3 dozen cupcakes for the party. How many cupcakes did she bake altogether?

_____

2 At the party supply store Ms. Porter bought 4 packs of balloons. Each pack had 12 balloons. How many balloons did she buy?

_____

3 There will be 11 children at the party. Ms. Porter made favor bags for each guest. She decorated each bag with 6 stickers. She had bought 7 dozen stickers. How many stickers does she have left?

_____

4 Ms. Porter decided to have a piñata at the party. She filled the piñata with candy. When the piñata breaks, each guest will be able to get at least 4 pieces of candy. What is the least number of pieces of candy in the piñata?

_____

5 To decorate for the party, Ms. Porter hung streamers. She bought 2 rolls of pink streamers and 3 rolls of purple streamers. Each roll was 12 feet long. She has 5 feet of streamers left. How many feet did she use?

_____

Solve for n.

6 $11 \times 9 = n$ _____   **7** $6 \times 12 = n$ _____   **8** $n = 7 \times 11$ _____

9 $12 \times 8 = n$ _____   **10** $n = 10 \times 11$ _____   **11** $12 \times 5 = n$ _____

12 $n \times n = 144$ _____   **13** $11 \times 8 = n$ _____   **14** $2 \times 12 = n$ _____

Name _____ **Date** _____

Estimating Area

Diego is thinking of a rectangle. He says, "It is at least 3 centimeters long but no more than 4 centimeters long. It is at least 1 centimeter wide but no more than 2 centimeters wide."

Which of these rectangles can be the one Diego is thinking of?

Write *yes* or *no* for each one. Then use a centimeter ruler to measure.

1

_____

2

_____

3

_____

Complete the following table about children's play areas. **Draw diagrams to help you.**

Play Area	Length (meters)		Width (meters)		Area (square meters)	
	At least	No more than	At least	No more than	At least	No more than
4 Mieko's	8	9	7	8	56	
5 Chelsea's	7	8	5	6		
6 Tyler's	6	7	5	6		
7 Desiree's	5	6	8	9		
8 Zachary's	8	9	5	7		

Real Math • Grade 4 • *Practice*

Name _____ Date _____

Finding Missing Factors

Solve each problem.

1 Jeffrey earns $5 each time he cleans his grandmother's house. How many times will he have to clean his grandmother's house to earn the $25 he needs for a new basketball?

_____

2 Ava rides her mountain bike 4 times each week. She rode the same distance each day she rode last week. She rode a total of 24 miles. How many miles did she ride each day?

_____

3 There are 4 quarters in a football game. The total playing time of a football game is 32 minutes. How many minutes long is each quarter?

_____

4 A baker uses a recipe that calls for 3 cups of flour to make 1 loaf of banana bread. The baker buys extra large bags of flour that contain 34 cups of flour. Is 1 bag of flour enough to make 12 loaves? Explain.

_____

_____

Solve for *n*.

5 $8 \times n = 48$ _____   **6** $7 \times n = 56$ _____   **7** $6 \times n = 42$ _____

8 $4 \times n = 28$ _____   **9** $5 \times n = 50$ _____   **10** $4 \times n = 12$ _____

11 $9 \times n = 36$ _____   **12** $27 = 3 \times n$ _____   **13** $8 \times n = 72$ _____

14 $n \times 6 = 54$ _____   **15** $64 = 8 \times n$ _____   **16** $5 \times n = 20$ _____

17 $n \times 3 = 9$ _____   **18** $n \times 6 = 18$ _____   **19** $45 = 5 \times n$ _____

20 $8 \times n = 32$ _____   **21** $5 \times n = 35$ _____   **22** $n \times 4 = 24$ _____

Multiplication and Division

Solve each problem.

1 Brianna earns $6 for each hour of work. She earned $48 today. How many hours did she work?

_____

2 When Brianna works overtime, she has to work only 4 hours to make $36. How much does she earn each hour working overtime?

_____

3 Last week Brianna worked 10 hours each day. She worked 4 days earning her regular pay and 1 day earning overtime pay. How much did she earn last week?

_____

Divide to solve for *n*.

4 $20 \div 4 = n$ _____

5 $27 \div 3 = n$ _____

6 $28 \div 4 = n$ _____

7 $16 \div 8 = n$ _____

8 $n = 36 \div 6$ _____

9 $72 \div 9 = n$ _____

10 $9 \div 1 = n$ _____

11 $42 \div 7 = n$ _____

12 $40 \div 4 = n$ _____

13 $n = 24 \div 6$ _____

14 $n = 27 \div 9$ _____

15 $30 \div 3 = n$ _____

16 $63 \div 7 = n$ _____

17 $n = 56 \div 8$ _____

18 $n = 25 \div 5$ _____

Find each quotient.

19 $1 \overline{)10}$

20 $4 \overline{)28}$

21 $7 \overline{)21}$

22 $6 \overline{)30}$

23 $7 \overline{)35}$

24 $2 \overline{)18}$

25 $8 \overline{)24}$

26 $3 \overline{)30}$

27 $3 \overline{)21}$

28 $6 \overline{)36}$

29 $4 \overline{)32}$

30 $6 \overline{)42}$

LESSON 3.11

Division with Remainders

Divide. Watch for remainders.

1 5)30 **2** 7)20 **3** 4)21 **4** 8)56

5 6)28 **6** 3)26 **7** 5)55 **8** 4)42

9 2)19 **10** 7)56 **11** 9)67 **12** 8)36

13 4)14 **14** 5)38 **15** 8)64 **16** 10)92

Read the information. Then answer the following questions.

Alan asked three of his friends if they wanted to start a recycling program in their neighborhood. The 4 boys collected aluminum cans, plastic, and paper from the neighborhood to take to the recycling center. Sometimes their neighbors paid the boys for taking away their recyclables. At the end of the week they divided all their money equally.

17 The first week they got $28 for their items at the recycling center. The neighbors also paid them $4.

   a. How much money did they earn altogether? _____

   b. How much should each boy get? _____

18 The second week they got $30 at the recycling center. They also received $6 from the neighbors.

   a. How much money did they earn altogether? _____

   b. How much should each boy get? _____

19 In the third week they got $25 at the recycling center and $5 in cash from their neighbors.

   a. How much money did they earn altogether? _____

   b. If they only have one-dollar bills, how many one-dollar bills should each boy get? _____

   c. How much money is leftover? _____

Name _____ **Date** _____

Common Multiples and Common Factors

Find the first three common multiples of each pair of numbers.

1 4 and 10 _____, _____, _____   **2** 3 and 5 _____, _____, _____

3 8 and 10 _____, _____, _____   **4** 2 and 7 _____, _____, _____

5 6 and 8 _____, _____, _____   **6** 5 and 8 _____, _____, _____

7 4 and 12 _____, _____, _____   **8** 7 and 9 _____, _____, _____

List the factors for each number. Then write each number pair's greatest common factor.

9 Factors of 6: _____   Factors of 12: _____

   Greatest common factor of 6 and 12: _____

10 Factors of 8: _____   Factors of 10: _____

   Greatest common factor of 8 and 10: _____

11 Factors of 3: _____   Factors of 9: _____

   Greatest common factor of 3 and 9: _____

12 Factors of 15: _____   Factors of 20: _____

   Greatest common factor of 15 and 20: _____

Answer the following question.

13 What is the largest whole number that divides exactly into both 7 and 12?

_____

Name _____ Date _____

Parentheses

Solve for *n*. Watch the parentheses and the signs.

1 $32 \div (8 \div 2) = n$ ___

2 $(18 - 3) \div 5 = n$ ___

3 $(32 \div 8) \div 2 = n$ ___

4 $18 - (5 + 5) = n$ ___

5 $28 - (4 \div 4) = n$ ___

6 $(48 \div 8) \div 2 = n$ ___

7 $(28 - 4) \div 4 = n$ ___

8 $(3 \times 2) \times 3 = n$ ___

9 $48 \div (8 \div 2) = n$ ___

Solve each problem below in two different ways. Put parentheses around the first pair of numbers before solving. Then put the parentheses around the second pair and solve.

10 $18 - 10 + 2 = n$

11 $3 \times 3 + 4 = n$

12 $5 \times 6 \div 3 = n$

_____ _____ _____

_____ _____ _____

13 $16 + 4 \times 2 = n$

14 $16 \div 4 \times 4 = n$

15 $16 - 3 \times 2 = n$

_____ _____ _____

_____ _____ _____

16 $3 \times 3 \times 4 = n$

17 $12 - 6 \div 3 = n$

18 $12 + 2 + 3 = n$

_____ _____ _____

_____ _____ _____

Solve each problem. Use each of the three rules you have learned for different methods of solving, and give each solution. Use parentheses to help.

19 $3 \times 4 \times 5 \times 6 = n$

20 $3 + 4 + 5 + 6 = n$

_____ _____ _____

_____ _____ _____

21 $3 \times 4 + 5 \times 6 = n$

22 $3 + 4 \times 5 + 6 = n$

_____ _____ _____

_____ _____ _____

Name _____   Date _____

Applying Math

Read each problem carefully, and then solve. Think about which operation to use.

1 Ramon had $8, and then he bought a fruit smoothie at the Juice Shack. He now has $5.75. How much money did he spend?

_____

2 Gillian is building a kite. She needs 2 pieces of balsa that are each 31 centimeters long and 1 piece that is 37 centimeters long. How much balsa does she need altogether?

_____

3 Ms. Hawk's garden is 8 meters long and 6 meters wide. What is its area?

_____

4 Nikole wants to invite 17 people to a party. Invitations are sold in packages of 6. How many packages of invitations should she buy?

_____

5 Nathan ordered 18 balloons for his sister Nikole's party. He chose 3 colors and ordered the same amount of each color. How many of each color did Nathan order?

_____

6 If 40 baseball cards are divided equally among 5 children, how many cards should each child get?

_____

7 The Abbot family and the Kwan family visited a Civil War battlefield last summer. Tickets cost $5 each. The Abbots bought 6 tickets, and the Kwans bought 4. How much did the two families spend on tickets altogether?

_____

Name _____ **Date** _____

Points on a Grid

Suppose the people of Square City agree to always give the street name first and then the avenue name second. Answer these questions.

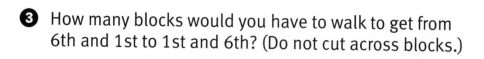

Square City

❶ The playground is at the corner of 6th and 1st. Where is the corner of 6th and 1st?

_____

❷ The bookstore is at the corner of 1st and 6th. Where is the corner of 1st and 6th?

_____

❸ How many blocks would you have to walk to get from 6th and 1st to 1st and 6th? (Do not cut across blocks.)

_____

Give the location of these points on the map of Square City. Remember to give the street name first and the avenue name second.

❹ C _____ ❺ D _____ ❻ E _____

Solve.

❼ Nikki lives at the corner of 9th Street and 9th Avenue (Point *G*). She needs to go to the toy store (Point *C*) to buy a gift for her cousin and then to the post office (Point *D*) to mail it. While she is out, she would like to stop at her friend's house (Point *E*) before returning home (Point *G*).

Describe the shortest route Nikki can take to visit all these places.

_____

_____

_____

_____

Name _____   Date _____

Coordinates

Solve this riddle by writing the correct letter for each of the coordinates from the graph.

❶ What did one math book say to the other?

____ , ____ ____ ____   ____ ____ ____
(1, 1) (12, 10) (4, 2)   (4, 10) (12, 8) (8, 9)

____ ____ ____ ____ ____ ____ ____ ____ .
(10, 2) (5, 8) (12, 8) (8, 1) (14, 9) (4, 2) (14, 3) (12, 5)

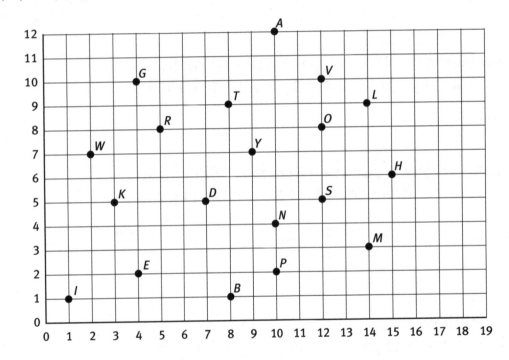

Use the grid above to answer these questions.

❷ List all of the points that are four steps away along straight lines from (6, 5).

_____

❸ Which point is closer to (9, 5)—the point (10, 4) or (4, 10)?

_____

❹ Which is closer to Point *S*—the point (17, 3) or (6, 2)?

_____

Name _____ Date _____

Lengths of Lines on a Grid

Solve.

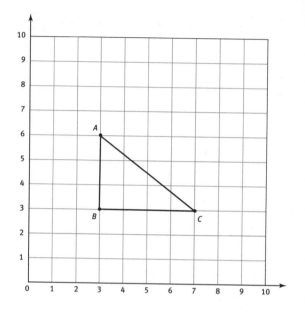

1 What is the length of Side *AB*?

_____

2 What is the length of Side *BC*?

_____

3 What is the area of a square with *AB* as its side?

_____

4 What is the area of a square with *BC* as its side?

_____

5 According to the Pythagorean Theorem, what would be the area of a square drawn on Side *AC*?

_____

6 What, then, is the length of Side *AC*?

_____

Answer the following questions.

The coordinates of *D, E,* and *F* are: *D* (7, 8), *E* (1, 8), and *F* (7, 0).

7 What is the length of Side *DE*? _____

8 What is the length of Side *DF*? _____

9 What is the length of Side *EF*? _____

Solve.

10 Using the Pythagorean Theorem, find the length of *N*.

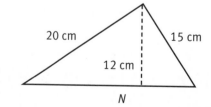

_____

Name _____   Date _____

Function Rules

Find a function rule for each set of arrow operations.

1 _____

10 →(?)→ 6
12 →(?)→ 8
14 →(?)→ 10
4 →(?)→ 0

2 _____

6 →(?)→ 30
2 →(?)→ 10
4 →(?)→ 20
5 →(?)→ 25

3 _____

20 →(?)→ 10
16 →(?)→ 8
12 →(?)→ 6
10 →(?)→ 5

4 _____

11 →(?)→ 17
8 →(?)→ 14
6 →(?)→ 12
24 →(?)→ 30

5 _____

0 →(?)→ 0
2 →(?)→ 4
6 →(?)→ 12
8 →(?)→ 16

6 _____

6 →(?)→ 6
4 →(?)→ 4
0 →(?)→ 0
50 →(?)→ 50

In each case, tell what y is.

7 14 →(+6)→ y _____

8 7 →(×5)→ y _____

9 10 →(÷2)→ y _____

10 24 →(−6)→ y _____

Solve.

11 One week Sharice worked 8 hours and earned $48. The next week her paycheck was $90 for working 15 hours. The third week she worked 12 hours.

a. What was the amount of Sharice's paycheck the third week? How do you know this?

_____

b. How much did Sharice earn altogether?

_____

Real Math • Grade 4 • *Practice*

Inverse Functions

Write the inverse operation.

1 +5 _____ **2** ×6 _____ **3** ÷3 _____ **4** +1 _____ **5** ×9 _____

6 −4 _____ **7** ÷7 _____ **8** ×2 _____ **9** +8 _____ **10** −3 _____

11 −2 _____ **12** +9 _____ **13** ÷5 _____ **14** ×1 _____ **15** +12 _____

Use inverse arrow operations, if they help you, to find the value of x.

16 x ──(+2)──▷ 9   _____   **17** x ──(×9)──▷ 72   _____

18 x ──(−12)──▷ 28   _____   **19** x ──(÷7)──▷ 9   _____

20 x ──(−3)──▷ 12   _____   **21** x ──(×8)──▷ 64   _____

22 x ──(+9)──▷ 19   _____   **23** x ──(+3)──▷ 27   _____

24 x ──(+10)──▷ 90   _____   **25** x ──(÷6)──▷ 6   _____

Solve.

26 Akira learned in science class that a cat can run as fast as 30 miles per hour, which is 2 times faster than a wild turkey. When he told his little sister what he had learned, she said, "So, if a cat can run 30 miles per hour, then a wild turkey can run 60 miles per hour."

   a. Is Akira's sister correct to say that a wild turkey can run 60 miles per hour?

   _____

   b. If not, explain why she is wrong and what she should have said.

   _____

   _____

Real Math • Grade 4 • *Practice*

Ordered Pairs

Copy the list of ordered pairs but replace the *x* or *y* with the correct number.

1 $x \longrightarrow \boxed{-4} \longrightarrow y$ (10, 6), (7, *y*), (*x*, 12), (10, *y*), (*x*, 0)

_____

Use the function rule to answer the question below. Find the value of *x* or *y* in each ordered pair. Then use the code to find what letter each number represents.

2 What is the largest member of the cat family? Use the function rule: $x \longrightarrow \boxed{+5} \longrightarrow y$

___   ___   ___   ___   ___   ___   ___   ___
(*x*, 11) (*x*, 14) (9, *y*) (5, *y*) (0, *y*) (4, *y*) (15, *y*) (8, *y*)

___   ___   ___   ___   ___
(2, *y*) (4, *y*) (12, *y*) (5, *y*) (*x*, 10)

X	O	L	M	R	S	T	H	I	E	W	Y	N
1	2	3	4	5	6	7	8	9	10	11	12	13

B	U	D	G	J	V	A	K	Z	C	P	Q	F
14	15	16	17	18	19	20	21	22	23	24	25	26

Ring the correct answer. In each problem, two of the answers are clearly wrong and one is correct.

3 309 + 617

a. 2,016

b. 326

c. 926

4 3,103 − 2,568

a. 535

b. 5,035

c. 135

5 1,647 + 9,353

a. 20,000

b. 8,000

c. 11,000

Name _____ **Date** _____

Function Rules and Ordered Pairs

Complete the tables. Solve for *x* or *y*.

1 $x \rightarrow \div 6 \rightarrow y$

x	*y*
36	
	5
24	
18	
	1

2 $x \rightarrow \times 4 \rightarrow y$

x	*y*
	20
	0
2	
8	
	28

3 $x \rightarrow -6 \rightarrow y$

x	*y*
20	
	24
40	
	44
60	

Solve.

4 Elizabeth is making pancakes for her friends, but her recipe makes only 14 pancakes. She needs to figure out how to modify her recipe so she will have at least 34 pancakes. Create a function machine and a set of ordered pairs to find out how many cups of pancake mix are needed to make at least 34 pancakes.

Pancake Recipe (makes 14 pancakes)
- 2 cups pancake mix
- 1 cup milk
- 2 eggs

_____

5 How many eggs would Elizabeth need to make at least 40 pancakes?

_____

$x \rightarrow \bigcirc \rightarrow y$

x	*y*

Name _____ Date _____

Graphing Ordered Pairs

Copy each set of ordered pairs but replace *x* or *y* with the correct number. Then graph each set of ordered pairs.

❶ *x* ──(+6)──▶ *y*   (4, *y*), (2, *y*), (5, *y*), (*x*, 13) _____

❷ *x* ──(×3)──▶ *y*   (1, *y*), (2, *y*), (4, *y*), (*x*, 9) _____

❸ *x* ──(−7)──▶ *y*   (10, *y*), (15, *y*), (*x*, 0), (*x*, 4) _____

❹ *x* ──(÷2)──▶ *y*   (14, *y*), (8, *y*), (*x*, 3), (*x*, 5) _____

❺ *x* ──(+10)──▶ *y*   (0, *y*), (*x*, 12), (4, *y*), (*x*, 16) _____

❻ *x* ──(×1)──▶ *y*   (1, *y*), (*x*, 3), (5, *y*), (*x*, 10) _____

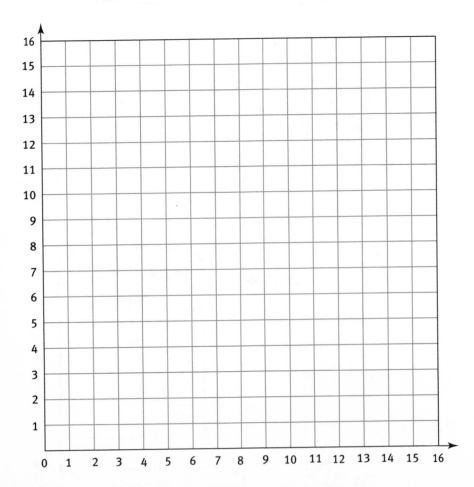

Real Math • Grade 4 • *Practice*

Name _____ **Date** _____

Identifying Scale

Solve.

Trevor opened up a savings account last year. The table shows his monthly account averages. He wants to make a graph to show the amount in his account each month.

1 What scale should Trevor use for the vertical axis on his graph if his graph paper is 25 squares tall?

_____

_____

2 Using this scale, how many increments in the *y* direction will be the amount for June?

_____

3 Using this scale, how many increments in the *y* direction will be the highest amount?

_____

4 Create a graph using the information in the chart.

Trevor's Savings Account	
January	$ 25
February	$ 35
March	$ 50
April	$ 40
May	$ 65
June	$ 75
July	$ 90
August	$ 60
September	$ 50
October	$ 60
November	$ 80
December	$100

Name _____ Date _____

Composite Function Rules

Find the value of *y*.

1 7 →(+2)→ *n* →(×5)→ *y* _____

2 30 →(÷6)→ *n* →(×7)→ *y* _____

3 12 →(−6)→ *n* →(÷3)→ *y* _____

4 18 →(−8)→ *n* →(+4)→ *y* _____

Solve.

Brian mows lawns to earn money. He charges $10 for each time he mows someone's lawn. For each new customer he charges a one-time start-up fee of $5.

5 Complete the table. Then graph the ordered pairs. Be mindful of the scale used in creating your graph. Connect the points on your graph.

6 Mr. Kish was a new customer this summer. Brian mowed his lawn 8 times. How much should Brian charge Mr. Kish when he sends his bill?

_____

7 Let's say that Brian mows 10 lawns.

   a. What is the most he could earn? _____

   b. What is the least he could earn? _____

8 If Brian earned $90, how many lawns did he mow and for how many customers? Try to find at least 3 possible solutions.

_____

_____

x →○→ *n* →○→ *y*

x	*y*
1	15
2	
3	
4	
5	

Name _____ **Date** _____

Inverses of Composite Functions

Find the value of *x*.

1 *x* →(×2)→ *n* →(×6)→ 36   *x* = _____

2 *x* →(×2)→ *n* →(+10)→ 16   *x* = _____

3 *x* →(+5)→ *n* →(÷2)→ 7   *x* = _____

4 *x* →(+5)→ *n* →(−9)→ 0   *x* = _____

5 *x* →(÷6)→ *n* →(×3)→ 27   *x* = _____

6 *x* →(÷4)→ *n* →(−2)→ 0   *x* = _____

7 *x* →(+5)→ *n* →(÷4)→ 5   *x* = _____

8 *x* →(−9)→ *n* →(×8)→ 72   *x* = _____

9 *x* →(÷10)→ *n* →(×8)→ 40   *x* = _____

10 *x* →(×7)→ *n* →(−20)→ 22   *x* = _____

Solve.

11 There are 3 students absent from Kia's ballet class. The students who are at class are divided into 4 groups of 3 students each.

a. How many students are at the ballet class today?

_____

b. How many students are in the class when no one is absent?

_____

12 Justin ordered some new CDs from an online music store. The store was selling the CDs for $9 each plus a shipping and handling charge of $6 for the order. He spent a total of $42.

a. What was Justin's cost before the shipping and handling fee?

_____

b. How many CDs did Justin order?

_____

Name _____ **Date** _____

Graphing Composite Functions

Use inverse operations to replace *x* with the correct number. Then graph each set of ordered pairs.

① *x* —(×3)→ *y*   (*x*, 6), (*x*, 15), (*x*, 30), (*x*, 0), (*x*, 24) _____

② *x* —(+6)→ *y*   (*x*, 10), (*x*, 16), (*x*, 26), (*x*, 6), (*x*, 24) _____

③ *x* —(−8)→ *y*   (*x*, 8), (*x*, 16), (*x*, 4), (*x*, 2), (*x*, 5) _____

④ *x* —(÷2)→ *y*   (*x*, 6), (*x*, 7), (*x*, 9), (*x*, 10), (*x*, 5) _____

⑤ *x* —(×2)→ *n* —(+6)→ *y*   (*x*, 8), (*x*, 10), (*x*, 12), (*x*, 20) _____

⑥ *x* —(÷4)→ *n* —(−4)→ *y*   (*x*, 2), (*x*, 4), (*x*, 1), (*x*, 3) _____

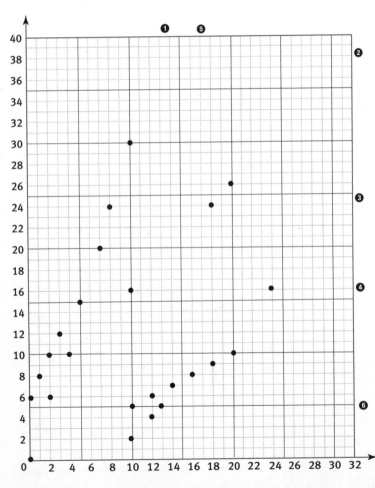

Real Math • Grade 4 • *Practice*

Name _____ **Date** _____

Graphing Functions

For each function rule, follow these steps.

Step 1 Find four ordered pairs of numbers.

Step 2 Graph the four points corresponding to these ordered pairs.

Step 3 Try to draw a straight line through all the points.

1 x ⟶ ÷1 ⟶ y **2** x ⟶ ÷2 ⟶ n ⟶ +4 ⟶ y

_____   _____

3 x ⟶ −5 ⟶ y **4** x ⟶ ×2 ⟶ y

_____   _____

Find a function rule for each graph.

5 A function rule is _____.

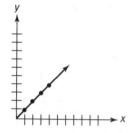

6 A function rule is _____.

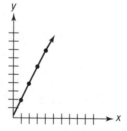

7 A function rule is _____.

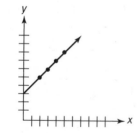

8 A function rule is _____.

Name _____ **Date** _____

Working with Graphs

Mei kept track of the amount of snow that fell each week this winter. She made two graphs to show the data she collected.

Answer the following questions.

1 Which graph better shows the total amount of snowfall?

_____

2 Which graph better shows the week in which the least amount of snow fell altogether?

_____

3 How much snow fell altogether?

_____

4 During which week did the most snow fall?

_____

5 Which graph did you use to answer Problem 2?

_____

6 Mei's birthday was during a week in which the snowfall amount was less than 6 inches but the total amount of snowfall was less than 24 inches. During which week was Mei's birthday?

_____

7 The temperature rose for three weeks in a row. During those weeks the snowfall amounts were low. In which weeks did the temperature rise?

_____

8 Look at Graph 2 for Weeks 6 and 7. What does the straight line tell you about the data?

_____

Graph 1—Weekly Snowfall Amounts

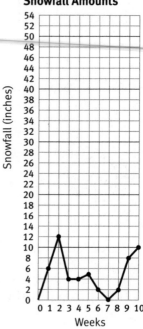

Graph 2—Running Total of Snowfall Amounts

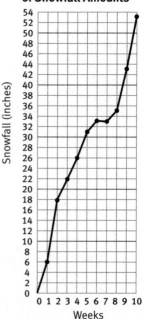

Name _____   Date _____

Misleading Graphs

Mr. Toshiko owns Computers Are Fun, Inc. The company has been in business since 1999. The table below shows the sales and profits for the company in its first seven years.

Year	Sales (Dollars)	Profits (Dollars)
1999	630,000	63,000
2000	510,000	48,000
2001	575,000	45,000
2002	620,000	57,000
2003	695,000	41,000
2004	750,000	69,000
2005	765,000	55,000

Answer the following questions using information in the table.

❶ In what year were the company's sales the greatest?

_____

❷ In what year did Mr. Toshiko earn the most profit?

_____

❸ Make line graphs of the sales and profits for Computers Are Fun, Inc. over the past seven years.

❹ Write about how fairly or unfairly the graphs represent the information.

_____

_____

❺ Make a graph for profits that shows the change in profit differently.

Name _____ **Date** _____

Multiplying by Powers of 10

Multiply.

1 100 × 5 _____

2 75 × 10,000 _____

3 89 × 10 _____

4 1,000 × 86 _____

5 10 × 10 _____

6 11 × 100 _____

7 6 × 1,000 _____

8 31 × 1,000 _____

9 100 × 60 _____

10 307 × 10 _____

11 41 × 10,000 _____

12 18 × 10,000 _____

13 1,000 × 8 _____

14 9 × 10,000 _____

15 12 × 10 _____

16 49 × 100 _____

17 824 × 100 _____

18 10,000 × 89 _____

19 64 × 10 _____

20 138 × 100 _____

21 100 × 10 _____

22 100 × 100 _____

Solve these problems.

23 The community theater is setting up chairs for a performance in the park. There will be 10 rows of chairs with 32 chairs in each row. They need to set up 40 more chairs. How many chairs have they set up so far?

_____

24 An architect made a scale model and a scale drawing of a building. On the drawing, the height of the building is 3 inches. The model is 10 times the size of the drawing and the actual building is 100 times the size of the model. What is the height of the actual building in inches?

_____

Name _____ Date _____

Converting Metric Units

Answer the following questions.

Remember

- 100 centimeters (cm) = 1 meter (m)
- 1,000 meters (m) = 1 kilometer (km)
- 1,000 grams (g) = 1 kilogram (kg)
- 1,000 milliliters (ml) = 1 liter (l)

---

1 How many centimeters are there in 6 meters? _____

2 How many meters are there in 6 kilometers? _____

3 How many centimeters are there in 62 meters? _____

4 How many meters are there in 62 kilometers? _____

5 How many milliliters are there in 6 liters? _____

6 Six dollars is worth how many cents? _____

7 How many cents is $62? _____

8 How many grams are there in 62 kilograms? _____

Ryan rode his bike all day. At the end of the day he said, "I rode 15 kilometers."

9 Do you think Ryan could have ridden 15 kilometers? _____

10 How far is that in centimeters? _____

11 During his bike ride, Ryan drank 3 liters of water. How many milliliters of water did Ryan drink? _____

12 Natalie biked 150 meters. Did she bike more than, less than, or the same distance as Ryan? Explain your answer.

_____

Name _____ Date _____

Multiplying by Multiples of 10

Multiply.

① 60 × 4 _____

② 40 × 500 _____

③ 60 × 60 _____

④ 80 × 4,000 _____

⑤ 6 × 40 _____

⑥ 7 × 700 _____

⑦ 90 × 300 _____

⑧ 8 × 3,000 _____

⑨ 60 × 40 _____

⑩ 3,000 × 2 _____

⑪ 20 × 50 _____

⑫ 60 × 7,000 _____

⑬ 600 × 400 _____

⑭ 50 × 50 _____

⑮ 8 × 900 _____

⑯ 800 × 50 _____

Write <, >, or = sign between the equations.

⑰ 20 × 500 _____ 5 × 2,000

⑱ 120 × 4 _____ 80 × 60

⑲ 600 × 40 _____ 30 × 80

⑳ 30 × 3,000 _____ 900 × 100

Solve these problems.

㉑ After buying lunch, Carter has 80¢ in change left over. He places this change in his piggy bank. If he does this every day, how much change will be in his piggy bank after 1 month (30 days)? Write your answer in cents and then in dollars. _____

㉒ If he continues this for 5 years (60 months), how much money will he have saved? _____

Name _____ Date _____

Practice with Multiples of 10

Answer the following questions.

> **Remember**
>
> - 100 centimeters in 1 meter
> - 1,000 meters in 1 kilometer
> - 100 cents in 1 dollar
> - 10 dimes in 1 dollar
>
> - 10 decimeters in 1 meter
> - 100 millimeters in 1 decimeter
> - 1,000 grams in 1 kilogram
> - 1,000 milliliters in 1 liter

1 How many centimeters are there in 60 meters? _____

2 How many meters are there in 5 kilometers? _____

3 How many cents are there in $300? _____

4 How many dimes are there in $300? _____

Solve each problem.

5 A paper clip is 5 centimeters long. How many paper clips would there be in a chain as long as the Mississippi River, which is almost 4,000 kilometers long? _____

6 Brady walks at a rate of 80 meters per minute. How many meters will Brady have walked in 1 hour (60 minutes)? Is that more than 1 kilometer? _____

7 The local radio station is doing a coin drive to raise money for the community. They raised a total of $900.

   a. If all the coins are dimes, how many dimes did the radio station collect? _____

   b. If all the coins are nickels, how many nickels did the radio station collect? _____

Name _____ **Date** _____

Rounding and Approximating

Approximate to solve these problems. **Explain how you found each answer.**

1 Jordan has $10.00. He wants to buy 6 packs of baseball cards. The baseball cards cost $1.95 for each pack. Does Jordan have enough money to buy the cards?

_____

2 The students at Kennedy School need 2,000 soup labels to win a prize for the school. If each of 8 classes turns in about 200 labels, will they have enough to win the prize?

_____

3 Lian wants to buy 10 stickers that cost 10¢ each. Does she have enough money if she has exactly $1.10?

_____

4 Ricardo is 8 years old. Is he more than 4,000 days old?

_____

5 A baseball stadium has 50 rows of seats with 420 seats in a row. Will the stadium hold at least 20,000 people?

_____

6 At a basketball game, the concession stand sells bags of popcorn for $3.25. If 200 bags of popcorn are sold, will the concession stand make at least $500?

_____

7 Michael wants to buy 200 sheets of paper that are sold for $1.85 for a packet of 50 sheets. Will $6.00 cover the cost of 200 sheets?

_____

8 As part of a training program, Chandra wants to run 250 kilometers each month. Can she do this by running 6 kilometers each day?

_____

Name _____ Date _____

Approximating Answers

Ring the correct answer. In each problem, two of the answers are clearly wrong and one is correct.

1 42×16
a. 342
b. 672
c. 1,912

2 12×37
a. 444
b. 682
c. 1,042

3 84×61
a. 2,824
b. 5,124
c. 3,512

4 92×59
a. 6,418
b. 829
c. 5,428

5 702×48
a. 2,839
b. 432,142
c. 33,696

6 32×645
a. 20,640
b. 2,140
c. 200,740

7 602×501
a. 540,212
b. 301,602
c. 721,800

8 45×66
a. 1,970
b. 2,970
c. 3,970

9 $1,402 \times 76$
a. 1,841,200
b. 10,500
c. 106,552

10 $8,447 \times 340$
a. 2,871,980
b. 1,000,847
c. 4,200,500

11 122×333
a. 40,626
b. 562,406
c. 4,250

12 999×27
a. 4,842
b. 36,823
c. 26,973

Solve.

13 A playground is 32 meters wide and 76 meters long. Approximate the area of the playground. Your answer should be a range between two numbers.

_____

_____

Name _____ Date _____

Practice with Approximating

Approximate the answer to each problem by rounding the factors. Then use your calculator to find the exact product. Compare each approximation with each exact product.

	Factors	Approximation	Exact Product
1	756 × 763		
2	68,589 × 7,496		
3	550 × 85,000		
4	5,496 × 6,498		
5	12 × 4,490		
6	200,000 × 30,000		
7	39 × 41		
8	545 × 40,000		

Derek works at a grocery store. He earns $11 an hour and works 37 hours a week for 52 weeks per year.

9 Approximate Derek's yearly salary. _____

10 Compare your approximation with Derek's exact yearly salary.

_____

Real Math • Grade 4 • *Practice*

Name _____ Date _____

Multiplication: Two-Digit by One-Digit

Multiply. Use shortcuts when you can.
Check to see that your answers make sense.

1 35
× 3

2 19
× 7

3 46
× 5

4 62
× 2

5 74
× 3

6 76
× 8

7 91
× 9

8 59
× 6

9 73
× 5

10 43
× 4

11 85
× 4

12 78
× 8

13 98
× 3

14 49
× 6

15 56
× 7

16 80
× 3

17 45
× 7

18 93
× 4

19 88
× 2

20 89
× 3

21 $56 \times 9 =$ _____

22 $56 \times 5 =$ _____

23 $83 \times 2 =$ _____

24 $83 \times 8 =$ _____

25 $36 \times 9 =$ _____

26 $44 \times 4 =$ _____

27 $38 \times 7 =$ _____

28 $51 \times 8 =$ _____

29 $65 \times 5 =$ _____

30 When you multiply a two-digit number by a one-digit number, how many digits will the product have?

_____

31 Jennifer multiplied 2 numbers. The product was the greatest possible product of a two-digit and a one-digit number. Write the factors and the product.

_____

32 What is the least product of a two-digit number multiplied by a one-digit number greater than zero?

_____

Name _____ Date _____

Multiplication: Three-Digit by One-Digit

Multiply. Use shortcuts when you can.
Check to see that your answers make sense.

1 453
  × 8

2 400
  × 7

3 569
  × 5

4 436
  × 4

5 585
  × 7

6 568
  × 3

7 600
  × 8

8 382
  × 9

9   64
  × 6

10 821
  × 5

11 521
  × 2

12 400
  × 9

13 409
  × 7

14 401
  × 8

15 646
  × 7

16 458
  × 6

17   88
  × 5

18 986
  × 6

19 823
  × 4

20   85
  × 9

21 $300 \times 4 =$ _____

22 $697 \times 9 =$ _____

23 $689 \times 3 =$ _____

24 $751 \times 8 =$ _____

25 $193 \times 9 =$ _____

26 $505 \times 7 =$ _____

27 $469 \times 6 =$ _____

28 $469 \times 3 =$ _____

29 $200 \times 4 =$ _____

30 $844 \times 2 =$ _____

Name _____ Date _____

Multiplication Review

**Multiply. Use shortcuts when you can.
Check to see that your answers make sense.**

1 65
  $\times 7$

2 307
  $\times 3$

3 99
  $\times 4$

4 615
  $\times 8$

5 395
  $\times 2$

6 401
  $\times 6$

7 443
  $\times 7$

8 56
  $\times 3$

9 80
  $\times 8$

10 406
  $\times 9$

11 $56 \times 7 =$ _____

12 $15 \times 3 =$ _____

13 $92 \times 6 =$ _____

14 $720 \times 9 =$ _____

15 $107 \times 4 =$ _____

16 $321 \times 5 =$ _____

Solve these problems.

17 Sonja's car travels 23 miles on 1 gallon of gas. She has 9 gallons of gas left in her car. Does she have enough gas to make it to her grandmother's house that is 189 miles away? Explain your answer.

_____

18 Audrey baby-sits 14 hours per week. If she earns $6 an hour, how many weeks will it take her to save $500?

_____

19 Steve is researching new computers before he buys one. The table below shows the down payment, number of monthly payments, and payment amounts for the same computer at three different stores. Which store has the best buy?

_____

Computer Plus	Electronics N' More	PC Superstore
$99	$289	$75
9 months	8 months	7 months
$80/month	$75/month	$99/month

Real Math • Grade 4 • *Practice*

Name _____ **Date** _____

Exponents

Solve these problems.

1 $4^2 =$ _____ $2^4 =$ _____ **2** $5^2 =$ _____ $2^5 =$ _____

3 $6^3 =$ _____ $3^6 =$ _____ **4** $8^2 =$ _____ $2^8 =$ _____

5 $1^5 =$ _____ $5^1 =$ _____ **6** $7^3 =$ _____ $3^7 =$ _____

7 $3^2 =$ _____ $2^3 =$ _____ **8** $9^2 =$ _____ $2^9 =$ _____

Use exponents to complete these number sentences.

9 $5 \times 5 \times 5 \times 5 \times 5 \times 5 =$ _____ **10** $6 \times 6 \times 6 \times 6 \times 6 \times 6 =$ _____

11 $7 \times 7 \times 7 \times 7 \times 7 =$ _____ **12** $1 \times 1 \times 1 \times 1 \times 1 \times 1 \times 1 =$ _____

13 $9 \times 9 \times 9 =$ _____ **14** $10 \times 10 \times 10 =$ _____

15 $10 \times 10 \times 10 \times 10 =$ _____ **16** $10 \times 100 =$ _____

17 $8 \times 8 \times 8 \times 8 =$ _____ **18** $10 \times 10 \times 10 \times 10 \times 10 =$ _____

19 $3 \times 3 \times 3 \times 3 \times 3 \times 3 \times 3 \times 3 =$ ___ **20** $10,000 \times 10 =$ _____

21 $2 \times 2 \times 2 \times 2 \times 2 =$ _____ **22** $100,000 =$ _____

23 $4 \times 4 =$ _____ **24** $1,000,000 =$ _____

Real Math • Grade 4 • *Practice*

Name _____   Date _____

Perimeter and Area

Solve.

Caleb and Darian both built pens for their dogs. Caleb made
a pen 12 meters by 8 meters. Darian made a pen that is
15 meters by 6 meters.

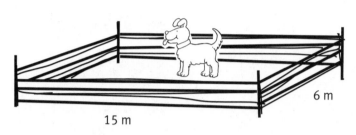
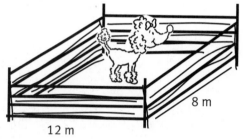

15 m   6 m   12 m   8 m

1 Who used more fencing to build his dog pen? _____

2 Whose dog has a larger play area? _____

3 Is it possible for Darian to build a pen with the same
perimeter and a greater area? Explain.

_____

_____

4 Using the same amount of fencing, what is the
greatest play area Caleb could build for his dog?
Draw a picture of the pen.

_____

Name _____ **Date** _____

Multiplication: Two-Digit by Two-Digit

Multiply. Use shortcuts when you can.

① 27
× 5

② 78
× 74

③ 18
× 20

④ 47
× 48

⑤ 43
× 40

⑥ 64
× 13

⑦ 91
× 8

⑧ 99
× 34

⑨ 95
× 36

⑩ 72
× 7

⑪ 28
× 46

⑫ 43
× 6

⑬ 21
× 89

⑭ 65
× 87

⑮ 80
× 80

⑯ 91
× 24

⑰ 53
× 27

⑱ 92
× 28

⑲ 69
× 15

⑳ 77
× 77

Check each answer below to see if it makes sense. Write *yes* or *no*. If the answer is wrong, write the correct answer.

㉑ 32
× 65
2,080
_____

㉒ 58
× 82
3,936
_____

㉓ 19
× 34
646
_____

㉔ 78
× 47
4,446
_____

Solve.

㉕ Ana is setting up chairs for an assembly. She knows that there are 525 students in the school. She sets the chairs in rows of 25. Will 20 rows be too few, just enough, or too many? Explain.

_____

Real Math • **Grade 4** • *Practice*

Applying Multiplication

Solve these problems.

1 Evie read 8 books for a summer reading program. Each book had about 95 pages. Did Evie read more than 750 pages? Explain.

_____

2 Steve bought 5 folders that cost 78¢ each, including tax. He gave the cashier $5.00. How much change did he get back?

_____

3 Ramón needs to buy 12 cans of soda, 3 boxes of crackers, and 2 packages of cheese. The soda costs 50¢ a can, the crackers cost $3 a box, and the cheese costs $2.10 a package. Will $20 be enough for him to buy what he needs? Explain.

_____

_____

4 Use the digits 2, 3, 7, and 9 to make a two-digit by two-digit multiplication problem with the greatest possible product.

_____ × _____ = _____

Balloons cost 24¢ each at Party Mart. At Balloons and More, you can buy a bag of 18 balloons for $3.78. You can have the balloons filled with helium for 5¢ each.

5 Colin wants to buy 32 balloons. How much will he have to pay at Party Mart? _____

6 How much will 32 balloons cost at Balloons and More? _____

7 How much will Colin save by buying his balloons at Balloons and More instead of Party Mart? _____

8 How much will it cost Colin to buy 32 balloons at Balloons and More and have them filled with helium? _____

Name _____ Date _____

Multiplication Practice

Multiply. Use shortcuts when you can.

1 32
× 46

2 84
× 21

3 45
× 16

4 58
× 37

5 83
× 94

6 72
× 30

7 67
× 56

8 25
× 38

9 49
× 29

10 60
× 50

11 29
× 18

12 79
× 64

13 26
× 21

14 17
× 15

15 64
× 63

Ring each correct answer. In each problem, two of the answers do not make sense and one is correct.

16 82
× 37

a. 3,034

b. 2,414

c. 30,304

17 53
× 53

a. 28,009

b. 2,806

c. 2,809

18 40
× 62

a. 248

b. 2,480

c. 240

19 74
× 45

a. 280

b. 3,330

c. 28,160

Solve each problem.

20 Mr. Landis drove from Pennsylvania to Georgia to visit relatives. The trip took 13 hours. Mr. Landis drove at an average speed of 55 miles per hour. How many miles was the trip? _____

21 The gas tank in Mr. Landis's car holds 12 gallons of gas. He can drive about 26 miles per gallon of gas. How many times did Mr. Landis have to fill his tank on his trip? (Hint: Use your answer from problem 20.) _____

Multiplication: Three-Digit by Two-Digit

Multiply. Use shortcuts when you can.
Check your answers to see if they make sense.

1 237
× 42

2 561
× 59

3 412
× 65

4 256
× 32

5 425
× 27

6 706
× 36

7 161
× 71

8 334
× 80

9 45
× 27

10 450
× 27

11 642
× 27

12 984
× 31

13 758
× 16

14 214
× 55

15 488
× 69

16 785
× 61

17 290
× 10

18 85
× 56

19 274
× 75

20 85
× 84

Each problem below gives the perimeter of a rectangle in inches and the area of the same rectangle in square inches. Give the dimensions of the rectangles that fit these numbers.

21 perimeter = 24
area = 32

length = _____

width = _____

22 perimeter = 18
area = 18

length = _____

width = _____

23 perimeter = 30
area = 50

length = _____

width = _____

24 perimeter = 32
area = 48

length = _____

width = _____

Solve.

25 The state theater has 104 rows of seats. Each row has 48 seats. Last night's show was sold out. How many tickets did the theater sell for the show?

_____

Converting Customary Units

Answer these questions.

1. How many ounces are there in 12 pounds? _____

2. How many pounds are there in 320 ounces? _____

3. How many inches are there in 5 feet? _____

4. How many inches are there in 5 yards? _____

5. How many feet are there in 11 yards? _____

6. How many feet are there in 96 inches? _____

7. How many pints are there in 8 gallons? _____

8. How many quarts are there in 8 gallons? _____

Solve these problems.

9. Mr. Wong bought a 6-foot long piece of wood. He cut off a piece that was 18 inches long. How many feet of wood does he have left? _____

10. Ella bought a gallon of milk. She used 12 ounces in a recipe, and then each of her 3 children drank an 8-ounce glass of milk. How many ounces of milk are left? _____

11. Mr. Finn bought $3\frac{1}{4}$ yards of fabric. Ms. Jackson bought 13 feet of fabric. Who bought more fabric? _____

12. Alison is making punch for a party. She used 1 pint of cranberry juice, 3 pints of orange juice, and 3 pints of ginger ale. Did she make a gallon of punch? _____

13. Sharia is making crispy rice treats. She bought two 2-pound bags of marshmallows. The recipe calls for 10 ounces of marshmallows. Sharia is making 4 batches of crispy rice treats. How many ounces of marshmallows will she have left? _____

Name _____ Date _____

Application of Multiplication

Complete the following computations by giving approximate and exact answers.

1 389×61 _____; _____

2 63×77 _____; _____

3 52×777 _____; _____

4 256×12 _____; _____

5 51×597 _____; _____

6 234×75 _____; _____

7 89×73 _____; _____

8 136×26 _____; _____

9 98×305 _____; _____

10 58×30 _____; _____

11 101×41 _____; _____

12 29×34 _____; _____

13 35×352 _____; _____

14 236×46 _____; _____

15 74×163 _____; _____

16 16×140 _____; _____

Solve these problems.

There are 13 players on the local baseball team. The baseball team decides to sell candy bars to raise money for new uniforms. Each box contains 24 candy bars. The baseball team must pay $12 for each box.

17 The baseball players decide to charge $1 per candy bar. If they sell 10 boxes, will they make a profit? How much? _____

18 Each uniform costs $78. How much money does the team need to raise to buy a new uniform for each player? _____

19 About how many boxes of candy bars must the team sell to earn enough to buy the uniforms? _____

20 If each player sells 120 candy bars, will the team raise enough money for the uniforms? Explain your answer.

LESSON 6.8

Multiplying Multidigit Numbers

Multiply. Use shortcuts when you can.

1 258
× 325

2 604
× 283

3 912
× 456

4 278
× 61

5 118
× 102

6 823
× 649

7 925
× 683

8 264
× 172

9 808
× 27

10 987
× 8

11 26
× 59

12 548
× 7

13 295
× 592

14 211
× 38

15 703
× 333

Below are four multiplication problems that have been solved. Two of them have errors. Ring the errors. Write *correct* below the problems that are correct.

16
```
    335
  ×  57
   2345
   1675
  19,095
```

17
```
    728
  × 627
   5096
   1456
   4368
  456,456
```

18
```
    419
  × 307
   2863
  12570
 128,563
```

19
```
    549
  × 126
   3295
   1088
    549
  69,074
```

Solve these problems.

20 The computer department ordered 20 boxes of paper. Each box contains 25 packages of paper. Each package of paper contains 250 sheets of paper. How many sheets of paper were ordered?

_____

21 José multiplied a three-digit number by a three-digit number. What is the greatest number of digits that the product could have? Give an example to support your answer.

_____

Real Math • Grade 4 • *Practice*

Name _____ Date _____

Using Multiplication

Solve these problems.

Jeremy travels about 65 miles in 1 hour on the highway. His truck goes about 18 miles on 1 gallon of gasoline. The gas tank holds 16 gallons. Liam travels about 55 miles in 1 hour on the highway. Liam's car goes about 23 miles on 1 gallon of gas, and her gas tank holds 12 gallons.

1 About how far can Jeremy travel on 1 full tank of gasoline?

_____

2 About how far can Jeremy travel in 4 hours on the highway?

_____

3 About how far can Liam travel on 1 full tank of gasoline?

_____

4 Who travels farther on 1 full tank, Jeremy or Liam? About how much farther?

_____

5 Who travels farther in 4 hours, Jeremy or Liam? About how much farther?

_____

6 Will either Jeremy or Liam have to fill up their tanks during a 4-hour trip? Explain.

_____

Name _____ Date _____

Approximating Products

Paint has spilled on this page of mixed addition, subtraction, and multiplication problems. Ring the correct answer in each case.

1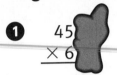
45
× 6

a. 28,086
b. 2,400
c. 260,345

2 43
+ 67

a. 43,204
b. 4,675
c. 44,543

3
× 0

a. 8,323
b. 1,405
c. 6,900

4 1036
− 1

a. 11,467
b. 9,120
c. 90,354

5 146
× 4

a. 6,278
b. 4,504
c. 1,464

6 85
− 45

a. 81,132
b. 90,132
c. 8,674

7 85
× 00

a. ,400
b. ,030
c. 0,000

8 73
+ 1 7

a. ,045
b. ,510
c. 361

9 8 6
− 3

a. 531
b. 1,106
c. 1,276

10 216
× 3

a. 5,806
b. 7,646
c. 76,464

Solve these problems.

11 Maria used 224 tiles to tile her kitchen. Each tile was 1 square foot. Give two possible dimensions for her kitchen.

_____

12 Maria's kitchen has a perimeter of 60 feet. What are the lengths of the sides of her kitchen? (Hint: Use the information from Problem 11.)

_____

13 The tiles Maria used for her kitchen cost $2.84 per tile. Estimate how much Maria spent on tiles. Explain.

_____

_____

Name _____ **Date** _____

Learning about Percentages

Look at each picture and the percentage given below it. Is the jar as full as each percentage says? Write *True* or *False*.

1

60% full

_____

2

75% full

_____

3

40% full

_____

4

90% full

_____

5

30% full

_____

6

85% full

_____

7

50% full

_____

8

25% full

_____

What part is shaded? Ring the correct percentage.

9

50% or 75%

10

5% or 20%

11

80% or 95%

Solve.

12 Is 100 percent of your class in school today? Explain.

_____

_____

_____

Percent Benchmarks

Solve the following exercises.

1 50% of 60 = _____   **2** 25% of 60 = _____   **3** 75% of 60 = _____

4 $\frac{1}{2}$ of 400 = _____   **5** $\frac{1}{4}$ of 400 = _____   **6** $\frac{3}{4}$ of 400 = _____

7 50% of 24 = _____   **8** 25% of 24 = _____   **9** 75% of 24 = _____

10 $\frac{1}{2}$ of 88 = _____   **11** $\frac{1}{4}$ of 88 = _____   **12** $\frac{3}{4}$ of 88 = _____

13 50% of 72 = _____   **14** 25% of 72 = _____   **15** 75% of 72 = _____

16 $\frac{1}{2}$ of 160 = _____   **17** $\frac{1}{4}$ of 160 = _____   **18** $\frac{3}{4}$ of 160 = _____

19 $\frac{1}{4}$ of 120 = _____   **20** 25% of 40 = _____   **21** $\frac{1}{2}$ of 70 = _____

22 $\frac{3}{4}$ of 100 = _____   **23** 75% of 20 = _____   **24** $\frac{3}{4}$ of 200 = _____

25 0% of 16 = _____   **26** 25% of 48 = _____   **27** $\frac{1}{4}$ of 48 = _____

Solve each problem. **Explain your answers.**

28 Selena wants a sweatshirt that usually sells for $18. Today the sweatshirt is on sale for 50% off. Selena has $10 with her. Can she afford to buy the sweatshirt?

_____

29 Tony saw a DVD player that usually sells for $80 but is on sale today for 25% off. Tony has saved $50 so far. Can he afford the DVD player today?

_____

Name _____ **Date** _____

Understanding $12\frac{1}{2}$% and $\frac{1}{8}$

Find $12\frac{1}{2}$% of the following quantities.

1 $12\frac{1}{2}$% of 48 = _____

2 $12\frac{1}{2}$% of 72 = _____

3 $12\frac{1}{2}$% of 88 = _____

4 $12\frac{1}{2}$% of 8 = _____

5 $12\frac{1}{2}$% of 800 = _____

6 $12\frac{1}{2}$% of 400 = _____

Give an equivalent fraction for each percentage or an equivalent percentage for each fraction.

7 25% of 1 = _____

8 $\frac{1}{2}$ of 1 = _____

9 75% of 1 = _____

10 50% of 2 = _____

11 $12\frac{1}{2}$% of 10 = _____

12 $\frac{3}{4}$ of 1 = _____

Look at the following problems, and answer *T* (true) or *F* (false). If the answer is false, write the correct percentage or fraction for the right side of the equation.

13 $\frac{1}{4}$ of 1 + 75% of 1 = 1 _____

14 50% of 1 + $\frac{1}{4}$ of 1 = 1 _____

15 $\frac{3}{4}$ of 1 + 25% of 1 = 2 _____

16 100% of 1 − 75% of 1 = $\frac{1}{2}$ _____

Solve the following exercises.

17 50% of 60 = _____

18 25% of 60 = _____

19 $\frac{1}{4}$ of 60 = _____

20 $\frac{1}{2}$ of 72 = _____

21 75% of 72 = _____

22 $12\frac{1}{2}$% of 72 = _____

Solve.

23 Yaniv took five spelling tests last month. The following percentages and fractions show how many of the words he spelled correctly each time: $\frac{3}{4}$, $37\frac{1}{2}$%, $\frac{1}{2}$, 100%, 70%. Change the fractions to percentages. Then arrange the scores in order from greatest to least.

_____

_____

Name _____ Date _____

Applying Percent Benchmarks

Complete the following tables.

❶ A mystery object is a rectangle 96 cm wide and 72 cm long. Complete the table about the object's width and length.

Benchmark Percent	Fraction	Width (in cm)	Length (in cm)
100%	$\frac{1}{1}$	96	
75%			
50%			
25%			18
$12\frac{1}{2}$%			

❷ Based on the data in the table above, which is probably the mystery object?

a. movie theater screen

b. magazine cover

c. classroom table top

Name _____ Date _____

Decimals and Stopwatches

Mieko's fourth-grade class timed her doing the sports activities found in the table below. Write the missing information.

	Activity	Fastest Time	Seconds	Centiseconds
1	50-yard dash	9.21		
2	Rope climb	15.97		
3	25 jumping jacks	17.06		

Solve.

4 Isabel and Haley wanted to see who could count faster. They timed each other as each counted by twos to 100. Isabel's time was 22 seconds and 16 centiseconds. It took Haley 22 and $\frac{1}{4}$ seconds.

a. Write both times as decimals. _____

b. Who counted faster? _____

c. Write both times as mixed numbers. _____

7.6

Name _____ Date _____

Adding and Subtracting Decimal Numbers

How large is the difference between the following:

1 0.44 and 0.18? _____ **2** 0.75 and 0.29? _____ **3** 0.60 and 0.04? _____

4 2.17 and 2.02? _____ **5** 5.93 and 5.51? _____ **6** 5.93 and 5.91? _____

7 6.89 and 6.09? _____ **8** 6.89 and 6.69? _____ **9** 6.89 and 6.88? _____

10 0.15 and 0.08? _____ **11** 0.71 and 0.17? _____ **12** 0.83 and 0.38? _____

13 1.61 and 1.01? _____ **14** 1.61 and 1.60? _____ **15** 3.33 and 3.01? _____

16 7.24 and 7.12? _____ **17** 9.07 and 9.00? _____ **18** 8.55 and 8.19? _____

19 3.65 and 0.05? _____ **20** 3.65 and 1.05? _____ **21** 0.68 and 0.06? _____

Solve each problem.

22 Jared bought 1.57 pounds of sliced ham. Was that more or less than a pound and a half? _____

23 Jill bought 0.71 pounds of salmon. Was that more or less than $\frac{3}{4}$ pound? _____

24 It took Selim 17.28 seconds to lace his hockey skates. It took Morgan 17.91 seconds to lace her skates. What is the difference in their times? Write your answer as both a fraction with a denominator of 100 and as a decimal.

_____

25 Kevin timed himself singing "The Star-Spangled Banner" as fast as he could. His three times were 23.06 seconds, 24.50 seconds, and 23.78 seconds.

 a. Which was his slowest time? _____

 b. Which was his fastest time? _____

 c. What was the difference between his slowest and fastest times? _____

74 Chapter 7 • *Introduction to Fractions, Decimals, and Percentages* **Real Math** • **Grade 4** • *Practice*

Copyright © SRA/McGraw-Hill.

Name _____ Date _____

Number Lines

Use each number line to answer the questions that follow it. Write your answers as decimals unless told otherwise.

①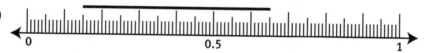

    a. Where does the bar start? _____

    b. Where does the bar end? _____

    c. What is the length of the bar? _____

    d. What percentage of the total
        length is the bar's length? _____

②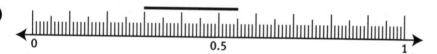

    a. Where does the bar start? _____

    b. Where does the bar end? _____

    c. What is the length of the bar? _____

    d. What percentage of the total
        length is the bar's length? _____

Solve.

③ Write any two decimals that fall between 0.25 and 0.50.
Then write each one as a fraction with a denominator
of 100.

    _____, _____, _____, _____, _____, _____

Name _____ Date _____

Understanding 10% and $\frac{1}{10}$

Apply what you have learned about 10% and $\frac{1}{10}$ to complete each statement.

1 $\frac{1}{10}$ of 400 = _____

2 10% of 500 = _____

3 10% of 600 = _____

4 $\frac{1}{10}$ of 700 = _____

5 $\frac{1}{10}$ of 800 = _____

6 10% of 900 = _____

7 10% of 60 = _____

8 $\frac{1}{10}$ of 70 = _____

Complete the following problems.

9 40% of 80 = _____

10 80% of 80 = _____

11 30% of 120 = _____

12 60% of 120 = _____

13 90% of 120 = _____

14 20% of 50 = _____

15 40% of 50 = _____

16 60% of 50 = _____

Solve each problem.

17 Sarah has been shopping for an MP3 player. One store has one she likes for 20% off the regular price of $90.

a. How much will Sarah save if she buys the MP3 player on sale?

_____

b. How much will Sarah pay for the MP3 player?

_____

18 Explain one way to find 90% of $40.

_____

Name _____ Date _____

Writing Appropriate Fractions

What fraction of each of the following figures has been shaded?

1 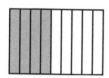 _____

2 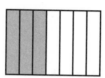 _____

3 _____

4 _____

5 _____

6 _____

7 _____

8 _____

9 _____

Which of these sentences are possible, and which are not possible? Explain how you know.

10 On Brandi's volleyball team, $\frac{1}{4}$ of the players are boys and $\frac{2}{3}$ of the players are girls.

_____

_____

11 Peter's mom baked an apple pie. Peter ate $\frac{1}{4}$ of the pie on Monday, $\frac{1}{4}$ of the pie on Tuesday, and $\frac{1}{4}$ of the pie on Wednesday.

_____

12 A carton of juice has 8 servings of 8 oz. of juice each. Ramón poured himself and three of his friends an 8 oz. glass of juice. Only $\frac{1}{4}$ of the juice is left in the carton.

_____

Name _____ **Date** _____

Fractions of Fractions

Complete the following to make correct statements. In each case we assume the fractions are fractions of the same whole.

1 $\frac{3}{4} \times \frac{1}{2} =$ _____

2 $\frac{1}{5} \times \frac{1}{2} =$ _____

3 $\frac{1}{4} \times \frac{1}{4} =$ _____

4 $\frac{1}{2} \times \frac{1}{2} =$ _____

5 $\frac{1}{4} \times \frac{1}{2} =$ _____

6 $\frac{3}{4} \times \frac{1}{3} =$ _____

7 $\frac{5}{6} \times \frac{2}{3} =$ _____

8 $\frac{2}{3} \times \frac{1}{3} =$ _____

9 $\frac{3}{8} \times \frac{2}{3} =$ _____

Replace the "?" to make each statement true.

10 $\frac{1}{3} \times \, ? = \frac{3}{15}$? = _____

11 $\frac{1}{4} \times \, ? = \frac{1}{12}$? = _____

12 $\frac{2}{3} \times \, ? = \frac{6}{12}$? = _____

13 $\frac{3}{8} \times \, ? = \frac{9}{32}$? = _____

14 $\frac{4}{5} \times \, ? = \frac{4}{10}$? = _____

15 $\frac{3}{10} \times \, ? = \frac{6}{30}$? = _____

Solve.

16 Mrs. Walker wants to make $\frac{2}{3}$ of a batch of cookies. For one batch she needs $\frac{3}{4}$ cup of sugar. How much sugar does she need to make $\frac{2}{3}$ of a batch?

_____

Name _____ **Date** _____

Fractions and Rational Numbers

Answer each question.

1 What is $\frac{1}{5}$ of a meter, in centimeters? _____

2 What decimal is equivalent to $\frac{1}{5}$? _____

3 What is $\frac{2}{5}$ of a meter, in centimeters? _____

4 What decimal is equivalent to $\frac{2}{5}$? _____

5 What is $\frac{4}{5}$ of a meter, in centimeters? _____

6 What decimal is equivalent to $\frac{4}{5}$? _____

Find either the decimal equivalent or an approximation to the nearest thousandth for each fraction or percentage.

7 $\frac{5}{6} = $ _____

8 $\frac{3}{10} = $ _____

9 $\frac{7}{9} = $ _____

10 $\frac{1}{3} = $ _____

11 $\frac{4}{9} = $ _____

12 $\frac{3}{3} = $ _____

13 $\frac{15}{100} = $ _____

14 $\frac{25}{100} = $ _____

15 $\frac{62}{100} = $ _____

16 $20\% = $ _____

17 $75\% = $ _____

18 $15\% = $ _____

19 $32\% = $ _____

20 $90\% = $ _____

21 $35\% = $ _____

Solve.

22 Evie has change in her pocket. She says the change totals $\frac{3}{4}$ of a dollar. How much is this written as dollars and cents?

_____

Probability

Answer the following questions.

1 If you roll a cube numbered 0–5, what is the probability that the number showing will be a 4?

_____

2 If you roll a cube numbered 0–5, what is the probability that the number showing will be greater than 3?

_____

3 If you roll a cube numbered 0–5, what is the probability that the number showing will be a 0?

_____

4 If you roll a cube numbered 0–5, what is the probability that the number showing will be less than 6?

_____

5 If you roll a cube numbered 0–5, what is the probability that the number showing will be even?

_____

6 If you spin this spinner, what is the probability that the pointer will land on a shaded section?

_____

Suppose you have a jar of 100 jelly beans. Of those, 30 are red, 45 are green, 15 are white, and 10 are yellow.

7 If you take a jelly bean from the jar without looking, what is the probability that the jelly bean you choose will be green?

_____

8 If you take a jelly bean from the jar without looking, what is the probability that the jelly bean you choose will be orange?

_____

9 If you take a jelly bean from the jar without looking, what is the probability that the jelly bean you choose will be white?

_____

10 If you take a jelly bean from the jar without looking, what is the probability that the jelly bean you choose will be red or yellow?

_____

Probability Experiments

Nicole and Toshiro are spinning a spinner that has the colors red, white, yellow, blue, green, and black. Toshiro wins if the color red or green is spun. Nicole wins if white, yellow, blue, or black is spun.

1 Who do you think will win more often? _____

2 What fraction of the time do you think Nicole will win? _____

3 What is Toshiro's probability of winning? _____

4 If they spin the spinner 6 times, how many times would you expect Nicole to win? _____

5 What is $\frac{4}{6}$ of 6? _____

6 Should you be surprised if Toshiro and Nicole won the same number of games after 6 spins? Explain your answer.

_____

Toshiro and Nicole played another game using a spinner and a **Number Cube**. The spinner was labeled with the numbers 0–5. The **Number Cube** was labeled with the numbers 1–6.

7 When spinning the spinner and rolling the **Number Cube,** how many different sums can be made? _____

8 How many different outcomes are possible? _____

9 How many different ways can you roll each sum?

a. 0 _____   **b.** 1 _____   **c.** 2 _____   **d.** 3 _____

e. 4 _____   **f.** 5 _____   **g.** 6 _____   **h.** 7 _____

i. 8 _____   **j.** 9 _____   **k.** 10 _____   **l.** 11 _____

10 What is the probability of rolling a sum of 6? _____

Name _____ **Date** _____

Applying Fractions

Answer the following questions.

1 There are 24 students in Sydney's class. Of those students, $\frac{3}{4}$ are right-handed. How many students in Sydney's class are left-handed? _____

2 The baseball glove that Jorge wants is on sale for $\frac{1}{3}$ off the original price. The original cost of the glove is $27. How much will Jorge save by buying the glove on sale? _____

3 There are 30 students in Mr. Turner's class, and $\frac{2}{5}$ of his students are girls. How many of his students are boys? _____

Solve the following exercises.

4 $\frac{1}{4}$ of 20 = _____

5 $\frac{3}{5}$ of 25 = _____

6 $\frac{2}{7}$ of 14 = _____

7 $\frac{2}{5}$ of 40 = _____

8 $\frac{5}{6}$ of 36 = _____

9 $\frac{2}{3}$ of 48 = _____

10 $\frac{1}{6}$ of 60 = _____

11 $\frac{5}{10}$ of 100 = _____

12 $\frac{3}{4}$ of 44 = _____

13 $\frac{3}{4}$ of 72 = _____

14 $\frac{3}{3}$ of 10 = _____

15 $\frac{5}{6}$ of 60 = _____

16 $\frac{2}{3}$ of 45 = _____

17 $\frac{2}{4}$ of 30 = _____

18 $\frac{1}{8}$ of 88 = _____

19 $\frac{4}{5}$ of 80 = _____

20 $\frac{2}{6}$ of 54 = _____

21 $\frac{3}{5}$ of 70 = _____

22 $\frac{1}{7}$ of 49 = _____

23 $\frac{1}{5}$ of 15 = _____

24 $\frac{4}{9}$ of 27 = _____

25 $\frac{1}{2}$ of 100 = _____

26 $\frac{1}{4}$ of 16 = _____

27 $\frac{1}{4}$ of 100 = _____

28 $\frac{1}{10}$ of 50 = _____

29 $\frac{1}{2}$ of 40 = _____

30 $\frac{2}{4}$ of 90 = _____

Name _____ Date _____

Probability and Fractions

Solve the following problems.

DeShawn and Riley have a jar filled with 8 red marbles,
2 white marbles, 10 blue marbles, and 4 green marbles.
One person is supposed to reach in and pick a marble
without looking. If the marble is red or white, DeShawn wins.
If the marble is blue or green, Riley wins.

❶ What is the probability that DeShawn will win? _____

❷ What is the probability that Riley will win? _____

❸ Suppose a marble is picked forty-eight times.

   a. How many times would you expect DeShawn to win? _____

   b. How many times would you expect Riley to win? _____

❹ How could you change the game so that each player
has an equal chance of winning?

_____

_____

What fraction of the circle is shaded?

❺ _____   ❻ _____   ❼ _____   ❽ _____

❾ Which is greater: $\frac{1}{4}$ of the circle or $\frac{3}{8}$ of the circle? _____

❿ Which is greater: $\frac{5}{8}$ of the circle or $\frac{1}{2}$ of the circle? _____

⓫ Which is greater: $\frac{1}{2}$ of the circle or $\frac{3}{6}$ of the circle? _____

⓬ Which is greater: $\frac{3}{4}$ of the circle or $\frac{7}{8}$ of the circle? _____

Name _____ **Date** _____

Equivalent Fractions

Convert each fraction to one that has a denominator of 18 and is equivalent.

1 $\frac{1}{3}$ _____

2 $\frac{1}{6}$ _____

3 $\frac{1}{2}$ _____

4 $\frac{1}{9}$ _____

5 $\frac{2}{3}$ _____

6 $\frac{4}{9}$ _____

Convert each fraction to one that has a denominator of 24 and is equivalent.

7 $\frac{1}{4}$ _____

8 $\frac{1}{3}$ _____

9 $\frac{1}{2}$ _____

10 $\frac{1}{6}$ _____

11 $\frac{3}{8}$ _____

12 $\frac{5}{6}$ _____

Solve the following problems.

13 Write a fraction with a denominator of 30 that is equivalent to $\frac{2}{5}$. In other words, $\frac{2}{5} = \frac{?}{30}$. _____

14 Write a fraction with a denominator of 48 that is equivalent to $\frac{2}{12}$. In other words, $\frac{2}{12} = \frac{?}{48}$. _____

15 Marisa's dad gave her 4 large jobs to do at their house. He gave her brother Oscar 8 small jobs to do. At the end of the week, Marisa had completed 3 of her jobs, and Oscar had done 6 of his. Oscar said he had done more of his work than Marisa had. Is this true? Explain how you know.

_____

Name _____ **Date** _____

Comparing Fractions

Write >, <, or =.

1 $\dfrac{3}{5}$ ☐ $\dfrac{4}{5}$ **2** $\dfrac{3}{4}$ ☐ $\dfrac{3}{8}$ **3** $\dfrac{2}{5}$ ☐ $\dfrac{6}{10}$ **4** $\dfrac{5}{6}$ ☐ $\dfrac{10}{12}$ **5** $\dfrac{1}{3}$ ☐ $\dfrac{3}{9}$

6 $\dfrac{2}{4}$ ☐ $\dfrac{1}{2}$ **7** $\dfrac{1}{2}$ ☐ $\dfrac{1}{3}$ **8** $\dfrac{1}{8}$ ☐ $\dfrac{1}{4}$ **9** $\dfrac{7}{12}$ ☐ $\dfrac{2}{3}$ **10** $\dfrac{7}{8}$ ☐ $\dfrac{3}{4}$

Solve the following problems.

11 Catherine made 20 pancakes. Her grandson Kiko ate 7 pancakes. Did Kiko eat more or less than $\dfrac{2}{5}$ of the pancakes? _____

12 Tyrone and Eli wanted to share 12 carrot sticks at lunch. If Eli eats 7 of the carrot sticks, will he have eaten more or less than $\dfrac{1}{2}$? _____

13 Kaitlin and Lilly both have MP3 players with 20 GB of memory. Kaitlin's MP3 player is $\dfrac{1}{3}$ full of songs, and Lilly's player is about $\dfrac{3}{12}$ full. Who has more room left on her MP3 player? _____

14 Desiree lives $\dfrac{3}{5}$ of a mile from the school. David lives $\dfrac{2}{3}$ of a mile from the school. Who lives farther from the school? _____

15 Jennifer typed $\dfrac{3}{5}$ of a 200-word essay for history. Ming typed $\dfrac{1}{4}$ of a 300-word essay for history.

a. Who typed more words? _____

b. How many more did she type? _____

Name _____ Date _____

Fractions Greater Than 1

Write a mixed number and an improper fraction to show how many.

1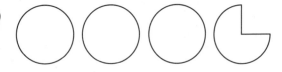

_____ apples

2

_____ circles

3

_____ pizzas

Write each mixed number as an improper fraction.

4 $3\frac{1}{5} =$ _____

5 $9\frac{2}{7} =$ _____

6 $5\frac{2}{3} =$ _____

7 $16\frac{1}{2} =$ _____

8 $11\frac{3}{4} =$ _____

9 $6\frac{7}{10} =$ _____

10 $1\frac{4}{9} =$ _____

11 $2\frac{7}{8} =$ _____

12 $6\frac{1}{3} =$ _____

Write each improper fraction as a mixed number or whole number.

13 $\frac{27}{6} =$ _____

14 $\frac{7}{2} =$ _____

15 $\frac{18}{9} =$ _____

16 $\frac{35}{7} =$ _____

17 $\frac{7}{4} =$ _____

18 $\frac{13}{6} =$ _____

Real Math • Grade 4 • *Practice*

Name _____ Date _____

Representing Fractions Greater Than 1

Fill in the missing information to make each statement true.

1 $\frac{8}{5} = \frac{\square}{20}$ _____

2 $\frac{3}{1} = \frac{12}{\square}$ _____

3 $\frac{13}{4} = \frac{65}{\square}$ _____

4 $\frac{14}{9} = \frac{\square}{27}$ _____

5 $\frac{5}{4} = \frac{\square}{100}$ _____

6 $3\frac{1}{2} = \frac{\square}{10}$ _____

Write the decimal equivalent of each improper fraction or mixed number.

7 $\frac{8}{5} =$ _____

8 $\frac{5}{2} =$ _____

9 $\frac{9}{8} =$ _____

10 $\frac{31}{10} =$ _____

11 $\frac{19}{4} =$ _____

12 $1\frac{3}{4} =$ _____

13 $3\frac{3}{10} =$ _____

14 $5\frac{9}{10} =$ _____

15 $7\frac{3}{8} =$ _____

Read each statement. Tell whether it is possible or not. If not, explain why.

16 Chad's dog weighs 500% the weight of his cat.

_____

17 Harry read 125% of the book.

_____

18 The height of the tree increased 160%.

_____

Solve.

19 A carton of milk holds 8 cups of milk. Julia used $2\frac{3}{4}$ cups of milk in a recipe. There are now $6\frac{1}{4}$ cups of milk left. Is this last sentence true? Explain how you know.

_____

Name _____ Date _____

Reading a Ruler

Estimate the length first. Then measure the length to the nearest $\frac{1}{8}$ inch.

❶

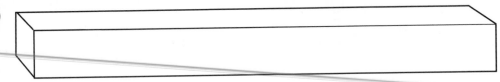

Estimate _____ Actual measure _____

❷

Estimate _____ Actual measure _____

❸

Estimate _____ Actual measure _____

❹

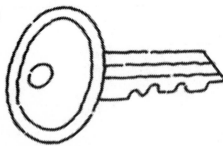

Estimate _____ Actual measure _____

❺

Estimate _____ Actual measure _____

Name _____ Date _____

Adding and Subtracting Measurements

Figure out the length of the third segment using the lengths of the two segments given below. After figuring out the answer, use a ruler to draw line segments of the correct length on your paper and check your answers.

1 $AB = \frac{3}{4}$ inch $BC = \frac{1}{8}$ inch $AC =$ _____

2 $AB = 1\frac{1}{4}$ inches $BC = \frac{1}{2}$ inch $AC =$ _____

3 $AB = \frac{3}{8}$ inch $BC = \frac{1}{4}$ inch $AC =$ _____

4 $AB = 1\frac{3}{8}$ inches $BC = 1\frac{1}{2}$ inches $AC =$ _____

Solve the following problems.

5 Jill placed a book that is $\frac{1}{2}$ inch thick on a book that is $\frac{3}{4}$ inch thick. How thick is the stack of two books?   _____

6 Corey swam $2\frac{1}{4}$ miles on Saturday and $2\frac{3}{4}$ miles on Sunday. How many miles did he swim in those two days?   _____

7 Simona bought $1\frac{1}{2}$ pounds of bananas and $2\frac{1}{4}$ pounds of apples. How many pounds of fruit did she buy altogether?   _____

8 Mr. Grossman had a 4-foot-long piece of wood. He cut off $\frac{1}{2}$ foot. How long was the remaining piece of wood?   _____

9 Adrienne made 5 gallons of punch for her birthday party. At the end of the party, $1\frac{7}{8}$ gallons were left. How much punch did Adrienne and her friends drink?   _____

10 Alex bought $\frac{3}{4}$ of a pound of turkey. He used $\frac{1}{8}$ of a pound to make a sandwich. How much turkey does Alex have left?   _____

Name _____ **Date** _____

Adding and Subtracting Fractions

Answer the following questions.

1 What is $\frac{1}{4} + \frac{3}{8}$? _____

2 How many eighths are shaded in the circle altogether? _____

3 What is $\frac{1}{2} + \frac{1}{5}$? _____

4 How many tenths are shaded in the circle altogether? _____

Solve. Watch the signs.

5 $\frac{1}{5} + \frac{2}{5} =$ _____

6 $\frac{4}{5} - \frac{2}{5} =$ _____

7 $\frac{1}{2} - \frac{1}{6} =$ _____

8 $\frac{3}{4} - \frac{1}{6} =$ _____

9 $\frac{1}{2} + \frac{3}{10} =$ _____

10 $\frac{4}{5} + \frac{1}{10} =$ _____

11 $\frac{7}{8} - \frac{1}{8} =$ _____

12 $\frac{1}{2} + \frac{3}{8} =$ _____

13 $\frac{1}{3} + \frac{1}{4} =$ _____

14 $\frac{2}{3} - \frac{7}{12} =$ _____

15 $\frac{1}{8} + \frac{3}{16} =$ _____

16 $\frac{11}{16} - \frac{1}{4} =$ _____

17 $\frac{8}{9} - \frac{2}{9} =$ _____

18 $\frac{3}{7} + \frac{4}{7} =$ _____

19 $\frac{7}{16} - \frac{3}{8} =$ _____

20 $\frac{1}{3} + \frac{1}{6} =$ _____

21 $\frac{5}{6} - \frac{1}{6} =$ _____

22 $\frac{1}{3} + \frac{5}{12} =$ _____

Name _____ Date _____

Adding Fractions Greater Than 1

Solve the following addition exercises. Write your answers as mixed numbers, fractions, or whole numbers.

1 $3\frac{1}{8} + \frac{7}{8} =$ _____

2 $4\frac{5}{32} + 3\frac{1}{8} =$ _____

3 $3 + 1\frac{5}{6} =$ _____

4 $\frac{1}{4} + \frac{1}{8} =$ _____

5 $5\frac{1}{2} + 2\frac{1}{4} =$ _____

6 $5\frac{10}{32} + 1\frac{3}{8} =$ _____

7 $9\frac{1}{16} + 4\frac{1}{4} =$ _____

8 $3\frac{2}{5} + 4\frac{3}{5} =$ _____

9 $6\frac{3}{4} + 2\frac{2}{8} =$ _____

10 $1\frac{5}{16} + 3\frac{3}{8} =$ _____

11 $4\frac{1}{3} + 1\frac{3}{8} =$ _____

12 $\frac{3}{2} + 4\frac{1}{2} =$ _____

Solve the following problems.

13 Chandra ran $2\frac{1}{2}$ miles on Monday and $3\frac{1}{8}$ miles on Tuesday. How far did she run over the two days? _____

14 Troy caught a $3\frac{5}{16}$-pound fish. His friend Dean caught a fish that weighed $5\frac{1}{8}$ pounds. How much did their two fish weigh together? _____

15 Mia is planning to bake blueberry muffins and pumpkin bread. The muffin recipe uses $2\frac{3}{4}$ cups of flour. The pumpkin bread recipe uses $3\frac{1}{8}$ cups of flour.

a. How much flour does Mia need for baking? _____

b. Mia only had $4\frac{3}{8}$ cups of flour, so she borrowed $2\frac{1}{2}$ cups of flour from her neighbor. Does she have enough flour? Explain. _____

Name _____ Date _____

Subtracting Fractions Greater Than 1

Solve the following addition and subtraction exercises.
Write your answers as mixed numbers, fractions, or whole
numbers.

1 $8\frac{6}{9} - 6\frac{3}{18} =$ _____

2 $7\frac{2}{3} - 4\frac{1}{6} =$ _____

3 $12\frac{2}{12} - 5\frac{1}{6} =$ _____

4 $3\frac{12}{16} - 1\frac{5}{8} =$ _____

5 $5\frac{7}{9} + 10\frac{4}{18} =$ _____

6 $7\frac{1}{2} - 3\frac{1}{4} =$ _____

7 $5\frac{2}{8} - 1\frac{1}{6} =$ _____

8 $9\frac{5}{6} + 2\frac{1}{12} =$ _____

9 $10\frac{1}{5} - 5\frac{1}{10} =$ _____

10 $6\frac{4}{5} + 7\frac{2}{10} =$ _____

11 $4\frac{3}{4} - 4\frac{1}{8} =$ _____

12 $6\frac{1}{2} + 3\frac{1}{3} =$ _____

13 $5\frac{3}{5} - 2\frac{1}{10} =$ _____

14 $4\frac{8}{32} - 2\frac{1}{8} =$ _____

Solve the following problems.

15 Ella's mother picked up 5 sub sandwiches for dinner
that were 1 ft long each. Ella and her sisters ate $2\frac{1}{4}$ of the
subs. How much is left over for the rest of the family?

_____

16 Mr. Krump gave Patrick and Lauren $6\frac{3}{4}$ ft of string to
use for their science fair project. If Patrick and Lauren
used $4\frac{5}{8}$ ft of string, how much is left over to return to
Mr. Krump?

_____

Real Math • Grade 4 • *Practice*

Name _____   **Date** _____

Parts of a Whole

Tell the value of the underlined digit in each of the following numbers.

① 0.8̲0 _____   **②** 0.19̲ _____   **③** 2̲.3 _____   **④** 30̲.51 _____

⑤ 11.24̲ _____   **⑥** 3̲.21 _____   **⑦** 33.07̲ _____   **⑧** 9.2̲8 _____

⑨ 1̲0.86 _____   **⑩** 57.2̲8 _____   **⑪** 6̲5.15 _____   **⑫** 1.04̲ _____

Write each number in standard decimal form.

⑬ 2 ones, 3 tenths, 6 hundredths   _____

⑭ 4 tens, 0 ones, 0 tenths, 3 hundredths   _____

⑮ 2 tens, 2 ones, 4 tenths, 9 hundredths   _____

⑯ 2 hundreds, 4 tens, 3 ones, 9 tenths, 2 hundredths   _____

⑰ 5 ones, 3 tenths, 7 hundredths   _____

⑱ 9 hundreds, 9 tens, 9 ones, 9 tenths   _____

Write each number in standard decimal form.

⑲ 50 + 8 + 0.6 + 0.05 _____   **⑳** 10 + 9 + 0.6 + 0.08 _____

㉑ 30 + 0 + 0.8 + 0.04 _____   **㉒** 40 + 3 + 0.3 + 0.04 _____

㉓ 90 + 0.7 + 0.02 _____   **㉔** 60 + 9 + 0.06 _____

㉕ A snail moves at a speed of three hundredths of a mile per hour. Write this value in standard form.

_____

Name _____ Date _____

Decimals and Fractions

Compare. Write <, >, or =.

1 0.5 ☐ 0.2   **2** 0.6 ☐ 0.8   **3** 0.020 ☐ 0.02   **4** 0.08 ☐ 0.03

5 0.09 ☐ 0.04   **6** 0.6 ☐ 0.06   **7** 0.087 ☐ 0.087   **8** 0.76 ☐ 0.8

9 0.6 ☐ 0.49   **10** 0.408 ☐ 0.532   **11** 0.268 ☐ 0.8   **12** 0.039 ☐ 0.53

Write each amount as a decimal.

13 7 dimes = _____

14 23 cents = _____

15 4 dimes and 6 cents = _____

16 9 cents = _____

Show what portion is shaded by writing a fraction and a decimal for each figure.

17

18

19

_____   _____   _____

Write each fraction as a decimal.

20 $\frac{18}{100}$ _____

21 $\frac{35}{100}$ _____

22 $\frac{3,428}{10,000}$ _____

23 $\frac{432}{1,000}$ _____

24 $\frac{5}{10}$ _____

25 $\frac{80}{100}$ _____

Real Math • Grade 4 • *Practice*

Name _____ Date _____

Comparing Decimals

Compare. Write <, >, or =.

1 6.8 ☐ 8.6

2 0.91 ☐ 0.893

3 42.6 ☐ 4.26

4 18.31 ☐ 18.310

5 462.5 ☐ 465.2

6 0.834 ☐ 8.1

7 7.76 ☐ 77.6

8 92.3 ☐ 8.14

9 10.62 ☐ 10.0

10 9.11 ☐ 8.10

11 8.3 ☐ 8.300

12 1.037 ☐ 1.03

Which of the following are possible? For those that are possible, write the coins needed. For those that are not possible, explain why.

13 Make 65 cents with eight coins.

_____

14 Make 45 cents with five coins.

_____

Solve these problems.

15 How many different coin combinations can you use to make 30 cents? Write 6 different combinations.

_____

_____

16 Caleb has 60 coins worth $1.00 altogether. What coins could Caleb have?

_____

Name _____ **Date** _____

Ordering Decimals

Rewrite each of the following sets of numbers from least to greatest with a $<$ symbol between each pair.

1 7.8, 8.1, 6.9, 7.79, 8.05, 6.945

_____

2 0.078, 0.081, 0.779, 0.0805, 0.06945, 0.69

_____

3 3.0, 4.002, 4.03, 3.98, 2.99967, 3.0056

_____

For each of the following pairs of numbers,

a. find a number that is exactly halfway between.

b. find a number that is closer to the lesser number.

c. find a number that is closer to the greater number.

d. write the three numbers in order from least to greatest.

4 8.4 and 9.3 **a.** _____ **b.** _____ **c.** _____ **d.** _____

5 7 and 8 **a.** _____ **b.** _____ **c.** _____ **d.** _____

Complete the following exercises.

6 For the following number line, say the number that should be placed where a letter is shown.

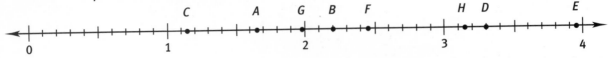

a. _____ **b.** _____ **c.** _____ **d.** _____

e. _____ **f.** _____ **g.** _____ **h.** _____

Name _____ **Date** _____

Rounding Decimals

Round to the nearest hundredth.

1 2.675 _____ **2** 4.1245 _____ **3** 0.0016 _____

4 62.329002 _____ **5** 6.213 _____ **6** 4.6173999 _____

7 2.462301 _____ **8** 1.817405006 _____ **9** 5.652 _____

Round to the nearest thousandth.

10 4.315621 _____ **11** 4.3156 _____ **12** 2.13590004 _____

13 0.00115 _____ **14** 6.3078 _____ **15** 1.9035003 _____

Solve these problems. Round the answers in a way that makes sense.

16 Ramon wants to buy a basketball that costs $16.99. He calculated the sales tax to be $1.10435. How much does the basketball cost, including tax?

_____

17 The fourth-grade classes are going on a field trip. The school needs buses to transport 648 people. Each bus can carry 52 people. Ms. Fritz divided 648 by 52 and got 12.461538. How many buses are needed for the trip?

_____

18 Jillian has taken 6 math tests. Her scores are 75, 86, 91, 78, 87, and 82. She found her average score to be 83.1666667. What score will Jillian get on her report card?

_____

Name _____ **Date** _____

Multiplying and Dividing by Powers of 10

Multiply.

1 6.3 × 100 = _____

2 0.41 × 10 = _____

3 42.61 × 100 = _____

4 903.1 × 10 = _____

5 28.418 × 100 = _____

6 10 × 3.034 = _____

Divide.

7 7.25 ÷ 10 = _____

8 0.5 ÷ 100 = _____

9 285 ÷ 100 = _____

10 11.2 ÷ 1,000 = _____

11 182.7 ÷ 1,000 = _____

12 3,907 ÷ 100 = _____

Solve. Watch the signs.

13 1,000 × 0.54 = _____

14 1,000 × 0.867 = _____

15 4.37 × 10 = _____

16 9.7 ÷ 10 = _____

17 6.86 ÷ 100 = _____

18 11.2 ÷ 100 = _____

Answer the following questions.

19 Dante saves $10.50 each week into his savings account. How much will be in Dante's account after 100 weeks?

_____

20 How many meters are there in 10.75 kilometers?

_____

Name _____ **Date** _____

Metric Units

Find the value of the missing number.

1 6 dm = _____ m

2 4 dm = _____ m

3 7 m = _____ dm

4 _____ dm = 9 m

5 1 m = _____ dm

6 6 m = _____ dm

7 70 m = _____ dm

8 21 cm = _____ m

9 9 cm = _____ m

10 0.42 m = _____ cm

11 _____ cm = 0.09 m

12 _____ cm = 1 m

13 4 mm = _____ m

14 127 mm = _____ m

15 37 mm = _____ m

16 _____ mm = 1 m

17 _____ mm = 0.810 m

18 0.7 m = _____ mm

19 2 km = _____ m

20 12 km = _____ m

21 60 m = _____ km

22 910 m = _____ km

Solve the following problems.

23 Each curtain panel in Jamila's front window measures 120 cm. Write this value in meters.

_____

24 Yesterday, Sheryl hiked 5.7 km. How many meters did she hike?

_____

Name _____ **Date** _____

Metric Measurements of Length

Write each measurement in meters.

1 5 m, 2 dm, 9 cm, 6 mm = _____ m

2 4 m, 3 dm, 2 cm, 7 mm = _____ m

3 7 cm, 3 mm, 2 m = _____ m

4 2 dm, 1 cm, 3 m = _____ m

5 1 m, 5 dm, 0 cm, 1 mm = _____ m

6 6 m, 1 mm, 2 cm = _____ m

7 6 cm, 9 dm, 3 m = _____ m

8 1 m, 9 mm, 7 dm = _____ m

Write each measurement in kilometers.

9 2 km, 350 m = _____ km

10 1 km, 700 m = _____ km

11 3 km, 30 m = _____ km

12 1400 m = _____ km

13 5 km, 100 m = _____ km

14 880 m = _____ km

Write the length of each object in millimeters and then in centimeters.

15

_____ mm   _____ cm

16

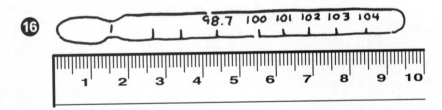

_____ mm   _____ cm

Real Math • Grade 4 • *Practice*

Name _____ Date _____

Adding and Subtracting Decimals

Add or subtract.

①
   4.12
 + 3.77
‾‾‾‾‾‾‾‾

②
   1.8
 + 2.72
‾‾‾‾‾‾‾‾

③
   6.84
 − 2.43
‾‾‾‾‾‾‾‾

④
  11.64
 + 7.30
‾‾‾‾‾‾‾‾

⑤
  19.07
 − 2.78
‾‾‾‾‾‾‾‾

⑥
   9.91
 − 2.38
‾‾‾‾‾‾‾‾

⑦
   4.03
 − 2.75
‾‾‾‾‾‾‾‾

⑧
   5.09
 + 4.91
‾‾‾‾‾‾‾‾

⑨
   2.4
 − 1.27
‾‾‾‾‾‾‾‾

⑩
   3.9
 − 1.06
‾‾‾‾‾‾‾‾

⑪
  10.3
 − 2.94
‾‾‾‾‾‾‾‾

⑫
   4.77
 + 3.53
‾‾‾‾‾‾‾‾

⑬
   6.07
 − 4.31
‾‾‾‾‾‾‾‾

⑭
   5.04
 + 3.19
‾‾‾‾‾‾‾‾

⑮
   7.98
 + 5.02
‾‾‾‾‾‾‾‾

Solve for *n*.

⑯ $1.72 + 3.14 = n$ $n =$ _____

⑰ $3.5 + 7.5 = n$ $n =$ _____

⑱ $6.29 − 4.24 = n$ $n =$ _____

⑲ $8.12 − 4.37 = n$ $n =$ _____

⑳ $9.65 + 2.7 = n$ $n =$ _____

㉑ $5.09 + 5.09 = n$ $n =$ _____

㉒ $3.04 − 2.07 = n$ $n =$ _____

㉓ $14.72 − 11.1 = n$ $n =$ _____

Solve the following problems.

㉔ Anudja swam 200 meters in 1 minute and 49.7 seconds. Catherine swam 200 meters in 2 minutes and 3.1 seconds. By how much did Anudja beat Catherine?

_____

㉕ Trevor swam 100 meters in 52.8 seconds. By how much time did he beat 1 minute?

_____

Name _____ Date _____

Using Decimals

Solve the following problems.

1 A 24-ounce jar of peanut butter costs $3.49. A 16-ounce jar of peanut butter costs $2.69. How much more does the 24-ounce jar cost than the 16-ounce jar?

_____

2 Ms. Williamson bought a skirt for $34.98 and a blouse for $19.99. How much did Ms. Williamson spend altogether?

_____

3 The DiFlorio family planned to drive 482 kilometers the first day of their vacation. They drove 135.6 kilometers before breakfast and 189.8 kilometers after breakfast. How much farther do they have to drive?

_____

4 On his first try, Bryce threw a softball 20.38 meters. On his second try he threw it 24.67 meters. How much farther did Bryce throw the softball on the second try?

_____

5 Marjan had $4,136.78 in her checking account. She then wrote a check for $23.45 and made a deposit of $135.75. How much is in Marjan's checking account now?

_____

6 Before Ms. Diaz went on a business trip, the odometer on her car showed 27,274.9 kilometers. After her trip it showed 28,109.4 kilometers. How many kilometers did she drive?

_____

Real Math • Grade 4 • *Practice*

LESSON 9.11

Balancing a Checkbook

Balance the following checkbooks.

1 Every month Jennifer gets a statement from her bank. The statement for May showed she had $321.46 in her account on May 4 and $159.36 at the end of the month. This did not agree with her records.

Look at Jennifer's checkbook. Did she make an error in her calculations? If she did, correct the error so that her records show the same balance at the end of the month as the statement shows.

CHECK	DATE	TRANSACTION	DEBIT		CREDIT		BALANCE	
							321	46
211	May 5	Roger's Jewelers	179	35			− 179	35
							142	11
212	May 11	Carson's Fashions	42	37			− 42	37
							184	48
213	May 17	Lander's Service Station	68	23			− 68	23
							116	25
214	May 22	Noteworthy News	22	15			− 22	15
							94	10
	May 24	deposit			150	00	+150	00
							244	10

2 On March 1st, Kia had $1,072.35 in her checking account. She deposits $535.72 into her checking account each Friday, and the month of March had four Fridays. In March, Kia wrote checks for the following amounts: $87.52, $27.76, $575, $52.80, $227.89, $47.13, $23.90, $30, and $456.70. She also made two withdrawals of $40 each from the ATM.

a. What was Kia's balance at the end of the month?

_____

b. Did Kia save money in March? If yes, how much?

_____

Name _____ **Date** _____

Multiplying by a Whole Number

Multiply.

1 1.8
    × 6

2 3.04
    × 5

3 7.18
    × 7

4 2.83
    × 3

5 2.19
    × 2

6 9.4
    × 8

7 4.12
    × 6

8 5.45
    × 5

9 6.9
    × 9

10 4.74
    × 4

11 4.003
    × 12

12 1.07
    × 24

13 8.14
    × 27

14 4.5
    × 16

15 2.4
    × 41

16 3.18 × 36 = _____

17 7.4 × 34 = _____

18 9.2 × 57 = _____

19 2.64 × 29 = _____

20 3.80 × 63 = _____

21 4.21 × 311 = _____

22 9.75 × 406 = _____

23 6.02 × 832 = _____

24 3.21 × 645 = _____

Solve the following problems.

25 Cameron wants to send out postcards to advertise his new restaurant. He plans to send 312 postcards. Each postcard costs 16¢. How much will the postcards cost?

_____

26 Janelle wants to buy 8 packages of party favors. Each package costs $2.79. Janelle has $20. Does she have enough money?

_____

27 Mr. Cortez buys 5 shirts for his children. Each shirt costs $12.75. He hands the clerk $70.00. How much change should he receive?

_____

Real Math • Grade 4 • *Practice*

Graphing and Applying Decimals

Complete the following exercises.

1 Use the function rule to complete the chart.

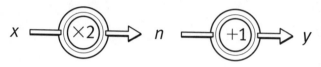

x	1.1	2.1	3.1	4.1	5.1
y	3.2	5.2			

2 Plot the points on the graph.

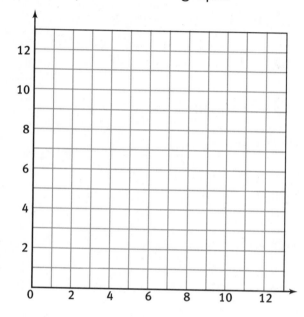

3 Do the five points seem to be on the same straight line?

_____

Choose the correct answer. In each problem two of the answers are clearly wrong, and one is correct.

4 $6 \times 2.3 =$ _____   **a.** 10.6   **b.** 13.8   **c.** 3.4

5 $5 \times 1.7 =$ _____   **a.** 4.5   **b.** 12.5   **c.** 8.5

6 $3.4 \times 8 =$ _____   **a.** 22.0   **b.** 27.2   **c.** 41.4

7 $0.9 \times 9 =$ _____   **a.** 8.1   **b.** 81   **c.** 0.81

8 $3 \times 0.43 =$ _____   **a.** 1.29   **b.** 3.71   **c.** 6.83

9 $0.02 \times 8 =$ _____   **a.** 0.16   **b.** 2.18   **c.** 16.02

Name _____ **Date** _____

Metric Units of Weight and Volume

Solve.

> **Remember**
> - 1 kg = 1000 g
> - 1 mL = 0.001 L
> - 1 km = 1,000 m
> - 1 L = 1,000 mL
> - 1 m = 1,000 mm

1 4 g = _____ kg

2 65 g = _____ kg

3 6 kg = _____ g

4 0.1 kg = _____ g

5 41 kg = _____ g

6 0.007 kg = _____ g

7 7 mL = _____ L

8 28 mL = _____ L

9 900 mL = _____ L

10 0.003 L = _____ mL

11 0.04 L = _____ mL

12 6 L = _____ mL

13 3,500 m = _____ km

14 0.3 km = _____ m

15 3 km = _____ m

16 70 m = _____ km

Answer the following questions.

17 The largest bird is the ostrich, with an average height of 2.55 m and a weight of 156 kg.

  a. What is the height of an ostrich in millimeters? _____

  b. What is the weight of an ostrich in grams? _____

18 A raccoon weighs 6.0 kg and measures 776.5 mm, and a gray fox weighs 5,200 g and measures 0.9625 m. Which animal is bigger? How much bigger? Explain.

_____

Name _____ Date _____

Cubic Centimeters

Find the volume of each box in cubic centimeters by figuring out how many cubes there are (assume all the cubes are 1 cubic centimeter).

 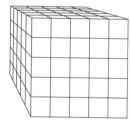

_____

 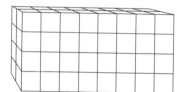

_____

 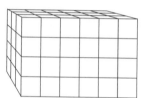

_____

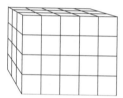

_____

Solve the following problems.

5 Which is a better value—one 1.8-liter bottle of water for $2.16 or three 500-mL bottles of water for $0.89 each?

_____

6 Miyoko has an empty jar that weighs 150 grams. She fills the jar with 1000 pennies. The jar now weighs 2.65 kg.

a. How much do the pennies in the jar weigh?

_____

b. On average, how much does each penny weigh?

_____

Name _____ **Date** _____

Lines

Identify each figure as a *point*, a *ray*, a *line segment*, a *straight line*, or an *angle*.

1 *R*   *S*
   •————•

_____

2 *K*   *L*
   •————•——→

_____

3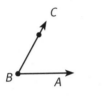

_____

4

_____

5

     •*M*

_____

6

_____

Answer the following questions.

7 Why is the figure to the right *not* an angle?

_____

8 Give two ways to name the figure to the right.

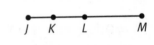

_____

9 Look at the figure to the right. How many line segments do you see? Be sure to count them all! Name them.

   J *K* *L*   *M*

_____

10 Look at the figures to the right. Which is shorter? Explain.

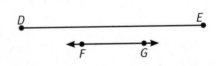

_____

Real Math • Grade 4 • *Practice*

Name _____ Date _____

Angles

Identify each angle as *acute*, *right*, *obtuse*, or *straight*.
Then use a protractor to find the measure of each angle in degrees. To make it easier to measure the angle with your protractor, you can extend its sides with your ruler.

1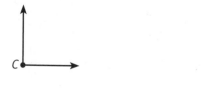

_____

Angle measure _____°

2

_____

Angle measure _____°

3

_____

Angle measure _____°

4

_____

Angle measure _____°

5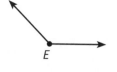

_____

Angle measure _____°

6

_____

Angle measure _____°

Solve these problems.

7 Look at the figure to the right to answer the following questions.

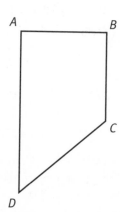

a. Which angle(s) is/are obtuse? _____

b. Which angle(s) is/are acute? _____

c. Which angle(s) is/are right? _____

8 Find the measure of each angle.

a. A _____

c. C _____

b. B _____

d. D _____

Real Math • Grade 4 • *Practice*

Name _____ Date _____

Parallel, Perpendicular, and Intersecting Lines

Tell whether the two lines are parallel, perpendicular, or neither in each case.

1

_____

2

_____

3

_____

4

_____

5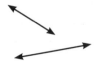

_____

6

_____

Solve these problems. You may draw pictures to help you.

7 What do you call two intersecting lines that meet at right angles?

_____

8 Identify five upper-case letters that have perpendicular lines.

_____

9 Identify five upper-case letters that have parallel lines?

_____

10 Identify five upper-case letters with intersecting lines that are neither parallel nor perpendicular.

_____

Name _____ Date _____

Quadrilaterals and Other Polygons

Identify each figure as a *pentagon*, a *hexagon*, an *octagon*, a *parallelogram*, a *rhombus*, a *rectangle*, a *square*, a *trapezoid*, or a *quadrilateral*. Choose the most descriptive term.

❶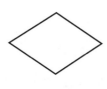

_____

❷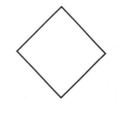

_____

❸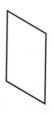

_____

❹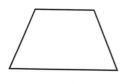

_____

❺

_____

❻

_____

❼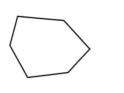

_____

❽

_____

❾

_____

Draw each figure as described in the space below.

❿ Draw a quadrilateral with exactly one pair of parallel sides.

⓫ Draw a hexagon with six equal sides.

Name _____ Date _____

Triangles

Identify each triangle as *equilateral*, *isosceles*, or *obtuse*.

①

_____

②

_____

③

_____

Draw a picture of each triangle in the space to the right.

④ Draw a right isosceles triangle.
Write *90°* in the right angle.

⑤ Draw an obtuse scalene triangle.
Write *O* in the obtuse angle.

⑥ Draw an acute scalene triangle.

Answer each question, and explain your thinking. Draw a sketch to help.

⑦ Can a triangle have two right angles?

_____

_____

⑧ Can the same triangle have an obtuse angle, a right
angle, and an acute angle?

_____

_____

Real Math • Grade 4 • *Practice*

Name _____ **Date** _____

Circles

Study the circle. Then identify its parts.

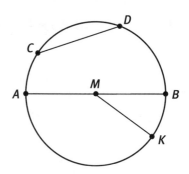

1 line segment *MB* _____

2 line segment *KM* _____

3 line segment *AB* _____

4 line segment *CD* _____

5 What is the point *M*? _____

6 What word names the length of the path around the circle?

_____

Draw the following parts in the circle with center *P*.

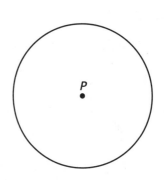

7 diameter *DE*

8 radius *PS*

9 chord *DL*

10 chord *SB*

Answer the questions. Explain your thinking.

11 How is a radius like a diameter? How is it different?

_____

_____

_____

12 Which part of a circle is always longer, diameter or radius?

_____

Name _____ **Date** _____

Congruence and Similarity

Use the figures below to answer the following questions.

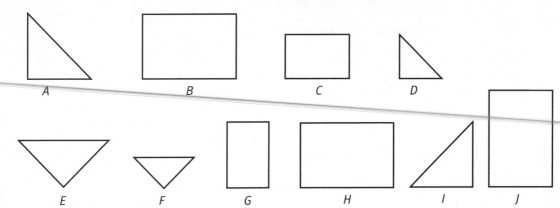

1 Which figures are congruent to *A*? _____

2 Which figures are similar to *A*? _____

3 Which figures are congruent to *B*? _____

4 Which figures are similar to *B*? _____

5 Which figures are similar to *I*? _____

Use a sheet of centimeter grid paper for the following problems.

6 Draw rectangle *ABCD* on the grid paper.
Make the long sides 5 cm long.
Make the short sides 2 cm long.

7 Calculate the area and perimeter of rectangle *ABCD*.

_____

8 Draw two different rectangles that are similar to *ABCD*.
Name each of them with four other letters.

_____

Name _____ Date _____

Rotation, Translation, and Reflection

Fill in the blanks to complete each statement. Write
rotation, *reflection*, or *translation*.

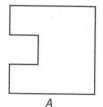 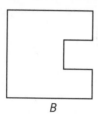 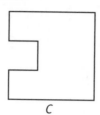 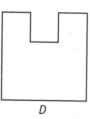

A   B   C   D

1 *B* is a _____ of *A*.

2 *C* is a _____ of *A*.

3 *D* is a _____ of *A*.

E   F   G   H

4 *F* is a _____ of *E*.

5 *G* is a _____ of *E*.

6 *H* is a _____ of *E*.

Answer each question by comparing the figures above.

7 Is *E* congruent to *C*? _____

8 Is *E* congruent to *F*? _____

9 What figures are congruent to *C*? _____

10 What figures are congruent to *E*? _____

Name _____ Date _____

Lines of Symmetry

Fill in the blanks to tell how many lines of symmetry there are in each.

1

_____

2

_____

3

_____

4

_____

5

_____

6

_____

Answer the following questions.

7 How many parallel lines are needed to divide a line segment into

   a. three equal parts? _____

   b. five equal parts? _____

   c. ten equal parts? _____

8 A line segment is 24 cm long.

   a. How many parallel lines are needed to divide it into eight equal parts?

   _____

   b. How long will each part be?

   _____

   c. What is $\frac{1}{8}$ of 24?

   _____

Name _____ Date _____

Space Figures

Name each space figure.

1
_____

2
_____

3
_____

4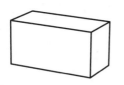
_____

5
_____

6
_____

What figures could you make from the following nets?

7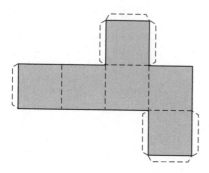

_____

8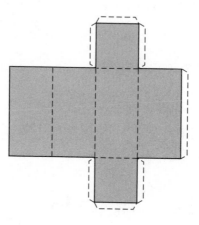

_____

Read and think about each statement. Then write *true* or *false*.

9 A sphere has a round base. _____

10 A pyramid has two congruent bases. _____

11 A can of soup is in the shape of a cylinder. _____

12 A football is a sphere. _____

Name _____ Date _____

The Five Regular Polyhedra

Name the regular polyhedra.

1

2

3

4

_____   _____   _____   _____

Complete each statement.

5 A(n) _____ is made up of polygons joined together to form a closed figure.

6 In a polyhedron, the places where two sides meet are

called _____.

7 _____ are the places where the corners of a polyhedron meet.

8 Every face in a polyhedron is a(n) _____.

9 A tetrahedron has _____ faces.

10 Every cube has eight _____.

11 The platonic solid known as the _____ has eight triangular faces.

Answer the following question.

12 Is a sphere a kind of regular polyhedron? Explain your reasoning.

_____

Name _____ Date _____

Pyramids and Prisms

Name each prism or pyramid.

1   _____

2   _____

3   _____

4 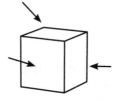 _____

Complete each statement.

5 The base of a pentagonal pyramid is a(n) _____.

6 A(n) _____ pyramid has a total of five faces.

7 A triangular prism has _____ vertices.

8 A pentagonal pyramid has ten _____.

The prism below is labeled *incorrectly*. Fix it by correctly labeling the prism on the right.

9 **a.**

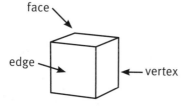

face

edge ← → vertex

b.

Answer the following questions.

10 What is the *fewest* number of faces a polyhedron can have? Explain.

_____

11 For any pyramid or prism, $V + F = E + 2$.
A *hexagonal* prism has 12 vertices and 8 faces.
How many edges does it have?

_____

Name _____ Date _____

Calculating Area

Find the area of each figure.

1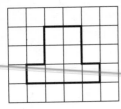

2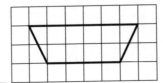

3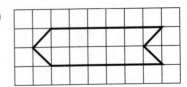

_____   _____   _____

Estimate the area of each figure.

4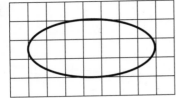

5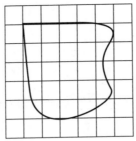

6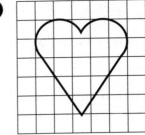

_____   _____   _____

Solve the following problem.

7 Look at the figure to the right. There are different ways to find its area. Find the area. Explain what you did.

_____

_____

8 cm
2 cm
6 cm   2 cm   4 cm

Name _____   **Date** _____

Measuring and Calculating Perimeter

Find the perimeter of each figure. Each square on the graph paper is one centimeter on a side.

❶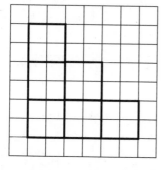

perimeter = _____

❷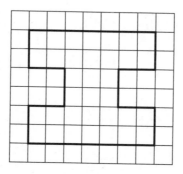

perimeter = _____

❸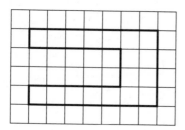

perimeter = _____

❹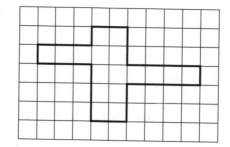

perimeter = _____

Solve the following problems.

❺ A square has an area of 64 square inches. What is its perimeter?

_____

❻ One side of a rectangle is 7 meters. Its area is 21 square meters. What is its perimeter?

_____

❼ A rectangular rug has an area of 16 square meters. It has a perimeter of 20 meters. What is its length? What is its width?

_____

❽ How many other rectangles can you draw that have a perimeter of 22 centimeters? Give the dimensions of each one.

_____

Name _____ Date _____

Dividing by a One-Digit Divisor

Divide. Record your steps on scrap paper. Watch for remainders.

1 $2\overline{)\$96}$ **2** $3\overline{)\$96}$ **3** $4\overline{)\$96}$ **4** $5\overline{)\$96}$ **5** $6\overline{)\$96}$

6 $5\overline{)\$555}$ **7** $8\overline{)\$888}$ **8** $3\overline{)\$666}$ **9** $2\overline{)\$666}$ **10** $3\overline{)\$967}$

11 $3\overline{)\$968}$ **12** $3\overline{)\$970}$ **13** $2\overline{)\$971}$ **14** $6\overline{)\$600}$ **15** $3\overline{)\$678}$

16 $6\overline{)\$678}$ **17** $3\overline{)\$912}$ **18** $5\overline{)\$6,572}$ **19** $4\overline{)\$9,876}$ **20** $7\overline{)\$907}$

Solve the following problems. Record your answers in the space given.

21 Ms. Lee's class held a bake sale. They earned a total of $428. They divided the money equally among 3 charities. How much money does each charity get?

Each charity gets _____.

The remainder is _____.

22 Mia wants a DVD player that costs $88. She can't afford to pay for it all at once. The store will let her pay it off over three months. Mia agrees to pay $40 the first month. Then she will pay the same amount each month after that. Figure out Mia's payment plan.

Month 1: _____

Month 2: _____

Month 3: _____

23 Suppose Mia paid the same amount each month for three months. How much would she pay per month? What would she do about the remainder? Explain a plan that works.

_____

_____

Name _____ **Date** _____

Division: Written Form

Divide. Keep your records in any of the ways you have learned.

1 2)‾468‾

2 3)‾963‾

3 4)‾1,689‾

4 4)‾1,609‾

5 6)‾424‾

6 6)‾4,242‾

7 7)‾424‾

8 7)‾4,242‾

9 5)‾378‾

10 5)‾3,784‾

11 5)‾37,849‾

12 8)‾99‾

13 8)‾351‾

14 9)‾1,881‾

15 3)‾1,881‾

16 2)‾98,765‾

Solve these problems. Record your answers in the space given.

17 Look carefully at Exercises 14 and 15 above. Describe how they are related. Explain how you could find the quotient for Exercise 15 without dividing but by using your answer from Exercise 14.

_____

_____

_____

18 Jared is saving for a skateboard that costs $67. His dad will give him $25 for the skateboard if he saves the rest. Jared can save $7 a week from his dog-walking pay. After how many weeks will Jared have enough for the skateboard? Explain your solution.

_____

_____

Name _____ Date _____

Checking Division

Divide. When there is a remainder, show it. Multiply to check your answers.

1 a. $7\overline{)246}$　　　**b.** Check: _____

2 a. $4\overline{)7,302}$　　**b.** Check: _____

3 a. $3\overline{)299}$　　　**b.** Check: _____

4 a. $3\overline{)399}$　　　**b.** Check: _____

Find the missing digit.

5
```
       662
  ■)4,634
     42
     43
     42
     14
     14
```
■ = ____

6
```
       968 R3
  4)3,■75
    36
    27
    24
    35
    32
     3
```
■ = ____

7
```
     8,■93
  9)74,637
    72
    26
    18
    83
    81
    27
    27
```
■ = ____

8
```
      753
  ■)753
    7
    5
    5
    3
    3
```
■ = ____

9
```
       985 R2
  3)2,■57
    27
    25
    24
    17
    15
     2
```
■ = ____

10
```
       893 R2
  5)■,467
    40
    46
    45
    17
    15
     2
```
■ = ____

11
```
       9,053 R3
  6)54,■21
    54
    32
    30
    21
    18
     3
```
■ = ____

12
```
     2,534
  ■)17,738
    14
    3 7
    3 5
    23
    21
    28
    28
```
■ = ____

Real Math • Grade 4 • *Practice*

Name _____ **Date** _____

Division: Short Form

Divide. Use whichever method you prefer. Check your answers by multiplying.

1 8)232   Check: Does 8 × _____ = 232?

2 5)750   Check: Does 5 × _____ = 750?

3 3)953   Check: Does 3 × _____ + _____ = 953?

4 7)604   Check: Does 7 × _____ + _____ = 604?

5 2)857   Check: Does 2 × _____ + _____ = 857?

Divide. Use shortcuts when you can.

6 4)716   **7** 4)719   **8** 9)515   **9** 6)139   **10** 3)139

11 2)1,686   **12** 2)686   **13** 7)58   **14** 7)581   **15** 5)922

16 6)328   **17** 6)330   **18** 1)13,945   **19** 1)37   **20** 2)817

21 8)800   **22** 8)400   **23** 9)66   **24** 5)456   **25** 5)4,560

Solve these problems.

26 Look at Exercises 18 and 19 above. Describe how they are related. Then explain how you can be sure that there won't be a remainder in either quotient.

_____

_____

27 Make up a division exercise that has an *odd* 3-digit quotient with *no* remainder. Use 2 as the divisor. Explain how you chose the dividend.

_____

_____

Name _____   Date _____

Division Patterns

Divide. Look for patterns that will help you find answers quickly.

1 $4\overline{)20}$ **2** $4\overline{)200}$ **3** $4\overline{)2,000}$ **4** $4\overline{)20,000}$ **5** $8\overline{)1,000}$

6 $8\overline{)2,000}$ **7** $8\overline{)4,000}$ **8** $9\overline{)100}$ **9** $9\overline{)1,000}$ **10** $9\overline{)10,000}$

11 $1\overline{)1,296}$ **12** $2\overline{)1,296}$ **13** $4\overline{)1,296}$ **14** $8\overline{)1,296}$ **15** $3\overline{)1,296}$

16 $6\overline{)1,296}$ **17** $2\overline{)148}$ **18** $2\overline{)74}$ **19** $3\overline{)927}$ **20** $5\overline{)927}$

Divide to solve for *n*.

21 $400 \div 8 = n$ **22** $400 \div 4 = n$ **23** $420 \div 7 = n$ **24** $4,200 \div 7 = n$

_____   _____   _____   _____

25 $3,618 \div 9 = n$ **26** $3,618 \div 6 = n$ **27** $3,618 \div 3 = n$ **28** $3,618 \div 2 = n$

_____   _____   _____   _____

Solve this problem.

29 Each summer, Steve and Maria take their 2 nieces to a theme park for the day. Tickets cost $12 for each adult and $7 for each child. Parking is $6. They bring $80. After they buy tickets and pay for parking, they agree to share the rest of the money equally. How much can each family member spend?

_____

_____

_____

Real Math • Grade 4 • *Practice*

Name _____ Date _____

Prime and Composite Numbers

Answer the following questions.
Remember: Prime numbers have exactly two factors.
Composite numbers have more than two factors.

1 List four prime numbers greater than 10.

_____

2 List all the composite numbers between 10 and 20.
How many did you find?

_____

List all the factors in order from least to greatest for each number.

3 12 _____   **4** 24 _____

5 10 _____   **6** 20 _____

7 37 _____   **8** 14 _____

How many factors does each number have?

9 13 _____   **10** 18 _____   **11** 11 _____

12 9 _____   **13** 16 _____   **14** 25 _____

15 5 _____   **16** 22 _____   **17** 30 _____

Think about each number. Write *composite* or *prime*.

18 2 _____   **19** 7 _____   **20** 9 _____

21 42 _____   **22** 29 _____   **23** 100 _____

Answer this question.

24 Why is 1 *not* a prime number?

_____

Name _____ **Date** _____

Finding Factors

Answer each question about divisibility by writing 2, 3, 5, and/or 11. Explain your answer. Watch out for prime numbers!

1 By what is 15 divisible? Why?

_____

_____

_____

2 By what is 232 divisible?

_____

_____

_____

3 By what is 88 divisible?

_____

_____

_____

4 By what is 588 divisible?

_____

_____

_____

5 By what is 370 divisible?

_____

_____

_____

6 By what is 43 divisible?

_____

_____

_____

7 By what is 730,215 divisible?

_____

_____

_____

8 By what is 111 divisible?

_____

_____

_____

9 Look back at Question 8. Why is 111 _not_ divisible by 11?

_____

Look at each set of numbers. Think about divisibility. Find the number that does not belong, and ring it. Explain your thinking.

10 22   24   77   66

_____

11 415   604   460   165

_____

Real Math • Grade 4 • _Practice_

Name _____ Date _____

Unit Cost

Solve these problems.

1 Matthew wants to buy some tea for his sister because she has a cold. A 6-ounce box costs $1.50 (150¢). A 9-ounce box costs $1.98 (198¢).

    a. How much does the tea cost per ounce in the 6-ounce box?

    _____

    b. How much does the tea cost per ounce in the 9-ounce box?

    _____

    c. Which is the better value?

    _____

2 Five bananas cost $1.10. What is the cost per banana?

_____

3 Three jars of strawberry jam cost $4.47. What is the cost per jar?

_____

Answer these questions.

4 A 6-slice pie costs $4.00. A 12-slice pie costs $6.99. Think about the unit cost. How much would you save buying the larger pie?

_____

5 Ian needs 15 pounds of clay. He can buy 1-pound cans of clay for 79¢, or 10-pound bags of clay at $6.99 each. Find the cheapest way for Ian to get the exact amount of clay he needs.

_____

Name _____ **Date** _____

Using Inverses

Use the graph to approximate how many dollars each of these items costs.

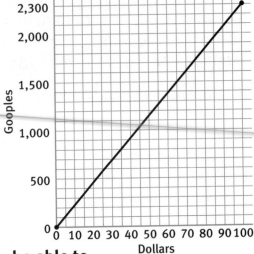

❶ CD Player—2,000 gooples: _____

❷ Backpack—820 gooples: _____

❸ Sunglasses—432 gooples: _____

❹ Book—98 gooples: _____

❺ Clock—1,220 gooples: _____

❻ Camera—1,760 gooples: _____

In each problem, find the value of *x* or *y*. You may be able to avoid some computing if you see a pattern.

❼ 6 ⟶ ×12 ⟹ *y* _____

❽ *x* ⟸ ÷12 ⟵ 72 _____

❾ 7 ⟶ ×244 ⟹ *y* _____

❿ *x* ⟸ ÷244 ⟵ 1,708 _____

⓫ 87 ⟶ ×10 ⟹ *y* _____

⓬ *x* ⟸ ÷10 ⟵ 870 _____

⓭ 9 ⟶ ×81 ⟹ *y* _____

⓮ *x* ⟸ ÷81 ⟵ 729 _____

⓯ 30 ⟶ ×65 ⟹ *y* _____

⓰ *x* ⟸ ÷65 ⟵ 1,950 _____

⓱ 52 ⟶ ×31 ⟹ *y* _____

⓲ *x* ⟸ ÷31 ⟵ 1,612 _____

⓳ 25 ⟶ ×75 ⟹ *y* _____

⓴ *x* ⟸ ÷75 ⟵ 1,875 _____

Real Math • Grade 4 • *Practice*

Name _____ **Date** _____

Estimating Quotients

The multiplication table gives products with 346. Use the products in the table to help you with the following exercises.

346 ×	2	3	4	5	6	7	8	9	10
=	692	1038	1384	1730	2076	2422	2768	3144	3460

For each division exercise below, estimate to find the best answer. Several answers are given but only one is close to the actual solution.

1 $6{,}288 \div 346 =$ **a.** 10   **b.** 12   **c.** 15   **d.** 18

2 $9{,}828 \div 346 =$ **a.** 2   **b.** 20   **c.** 28   **d.** 31

3 $13{,}498 \div 346 =$ **a.** 39   **b.** 35   **c.** 30   **d.** 28

4 $21{,}796 \div 346 =$ **a.** 70   **b.** 68   **c.** 66   **d.** 63

5 $24{,}520 \div 346 =$ **a.** 7   **b.** 70   **c.** 77   **d.** 707

6 $18{,}928 \div 346 =$ **a.** 47   **b.** 52   **c.** 59   **d.** 61

7 $73{,}892 \div 346 =$ **a.** 203   **b.** 302   **c.** 180   **d.** 401

8 $32{,}396 \div 346 =$ **a.** 98   **b.** 92   **c.** 89   **d.** 83

9 $17{,}108 \div 346 =$ **a.** 40   **b.** 47   **c.** 54   **d.** 62

10 $25{,}844 \div 346 =$ **a.** 66   **b.** 70   **c.** 73   **d.** 77

11 $38{,}584 \div 346 =$ **a.** 10   **b.** 95   **c.** 101   **d.** 106

12 $30{,}940 \div 346 =$ **a.** 80   **b.** 85   **c.** 89   **d.** 93

Solve this problem.

13 A gallon of paint covers about 400 square feet. Mr. Morano's garage has about 1850 square feet of wall area. If Mr. Morano wants to paint three coats, how much paint should he buy? Explain.

_____

Name _____ Date _____

Dividing by a Two-Digit Divisor

Divide. Give the answer to the nearest whole number.

❶ $223{,}262 \div 62$ **❷** $223{,}262 \div 31$ **❸** $12{,}987 \div 13$ **❹** $12{,}000 \div 13$

❺ $16{,}825 \div 673$ **❻** $16{,}152 \div 673$ **❼** $16{,}825 \div 25$ **❽** $110{,}745 \div 321$

❾ $55{,}728 \div 43$ **❿** $55{,}728 \div 438$ **⓫** $25{,}029 \div 81$ **⓬** $25{,}272 \div 81$

Choose the correct answer to the nearest whole number.
Use whatever method you wish.

⓭ $15{,}715 \div 35 =$	**a.** 449	**b.** 4,449	**c.** 494	**d.** 499
⓮ $15{,}715 \div 422 =$	**a.** 372	**b.** 375	**c.** 37	**d.** 35
⓯ $10{,}458 \div 14 =$	**a.** 77	**b.** 747	**c.** 774	**d.** 787
⓰ $10{,}458 \div 51 =$	**a.** 205	**b.** 215	**c.** 225	**d.** 235
⓱ $5{,}050 \div 5 =$	**a.** 101	**b.** 1,110	**c.** 1,010	**d.** 1,050
⓲ $5{,}050 \div 25 =$	**a.** 200	**b.** 202	**c.** 205	**d.** 2,002

Solve each problem.

⓳ The gas tank in Jamille's car holds 19 gallons. She usually drives 33 miles for each gallon of gas she buys. She is about to make a 550-mile trip in her car. Can she make the whole trip on one tank of gas? Explain.

_____

⓴ Next month, 384 students and adults are taking a bus trip to a computer museum. Each bus can seat 48 people. How many buses are needed for everyone?

_____

Real Math • Grade 4 • *Practice*

Name _____ **Date** _____

Applying Mathematics

Divide. Use shortcuts when you can.

❶ 7)‾560‾   **❷** 8)‾534‾   **❸** 4)‾100‾   **❹** 3)‾639‾   **❺** 2)‾946‾

❻ 5)‾450‾   **❼** 5)‾453‾   **❽** 9)‾631‾   **❾** 9)‾639‾   **❿** 4)‾287‾

Solve these problems.

⓫ It takes Valerie 5 minutes to iron a shirt. If she continues working at that pace, how many shirts can Valerie iron in two hours?

_____

⓬ Elan bought 7 cans of tomato soup. The clerk charged him a total of $2.73. How much did each can cost?

_____

⓭ Coach Jackson is putting together teams of 6 for a volleyball tournament. If 260 people sign up to play, how many teams can be formed?

_____

⓮ Nicole made 260 corn muffins this morning. She sells them in cartons of 6. How many complete cartons of corn muffins can she sell?

_____

⓯ Angel drove to visit his sister who lives 233 miles away. The trip took him 4 hours. About how many miles did Angel drive per hour?

_____

⓰ Pradip called his father in India last night. They spoke for 8 minutes. The call cost $2.16 (or 216 cents). How much did Pradip pay per minute?

_____

Name _____ Date _____

Finding Averages

Find the average of each set of numbers. Use shortcuts when you can.

1 4, 5, 6, 7, 8 _____

2 19, 20, 21, 22, 23 _____

3 15, 20, 25, 30, 35 _____

4 12, 15, 21, 19, 10, 7 _____

5 66, 66, 66, 66, 66, 66, 66 _____

6 1, 2, 3, 4, 5, 6, 7, 8, 9 _____

7 42, 61, 88, 24, 35 _____

8 4,634; 4,635; 4,636; 4,637; 4,638 _____

9 37, 47, 57, 67, 77, 87 _____

10 3,912; 3,914; 3,918; 3,920; 3,922; 3,916 _____

Solve these problems.

11 Hilary took 4 math tests. Her grades were 95, 83, 74, and 88.

a. What was her average grade for the 4 tests?

_____

b. Hilary took the fifth math test. Her average grade for the 5 tests was 85. What was her score on the fifth test?

_____

12 Namid bowled 3 games. His scores were 182, 173, and 209.

a. What was his average score for the 3 games?

_____

b. If Namid bowled a fourth game and scored 156, what would be his average score for the 4 games?

_____

Name _____ **Date** _____

Mean, Median, Mode, and Range

Find the mean, median, mode, and range.

1 10, 3, 17, 8, 12, 3, 9, 4, 6 _____

2 15, 10, 10, 11, 15, 9, 11, 15, 15, 13, 16, 16 _____

3 83, 54, 74, 64, 90, 56, 83 _____

4 9, 8, 3, 0, 8, 4, 10, 9, 6, 4, 9, 2 _____

Solve these problems.

5 Ms. Martinez recorded, in centimeters, the heights of the 14 students in the outdoor club. The heights are 121, 120, 121, 125, 130, 134, 160, 146, 140, 151, 150, 128, 141, and 123.

a. What is the average, or mean, height of the students?

_____

b. What is the median height of the students?

_____

c. What is the range of the heights?

_____

d. Sharice said that her height is the mode of the heights. How tall is Sharice?

_____

e. Of the mean, median, and mode of the heights, which do you think best represents the heights of the outdoor club students?

_____

f. A new student joined the club. His height is 135 centimeters. Explain how this changes the mean, median, mode, and range of the data.

_____

Name _____ Date _____

Using Mathematics

Solve these problems. Check to see that your answers make sense and that they fit the given situations.

1 Mr. Tindall is a real-estate agent in Clarksville. He recently sold seven houses. The prices of those houses were $185,000, $137,000, $162,000, $190,000, $115,000, $240,000, and $98,000.

   a. What was the average house price?

   _____

   b. How many houses were above-average in price?

   _____

   c. How many houses were below-average in price?

   _____

   d. Mr. Tindall used the median price of the houses he sold to determine the sale price for the Ramos's house. What is the sale price for the Ramos's house?

   _____

2 Mai-Lin recorded the daily high temperatures for the past six days. The high temperatures for each day were 57°F, 42°F, 32°F, 38°F, 62°F, and 42°F.

   a. What was the average high temperature for the week?

   _____

   b. What was the range of the temperatures?

   _____

   c. Over the last decade, the average high temperature for this time of year was 52°F. Would you say that this year is warmer than usual, cooler than usual, or about average?

   _____

Real Math • Grade 4 • *Practice*

Name _____ **Date** _____

Choosing Reasonable Answers

Choose the correct answer. Use shortcuts when you can.

1 5)318

   a. 6R3
   b. 16R3
   c. 33R6
   d. 63R3
   e. 66R6

2 40)200

   a. 5
   b. 50
   c. 80
   d. 500
   e. 800

3 2)26

   a. 10
   b. 11
   c. 12
   d. 13
   e. 14

4 6)474

   a. 8
   b. 24
   c. 79
   d. 84
   e. 108

5 3)345

   a. 15
   b. 19
   c. 59
   d. 115
   e. 150

6 7)100

   a. 9R2
   b. 11R3
   c. 14R2
   d. 40R3
   e. 98R2

7 9)162

   a. 16
   b. 17
   c. 18
   d. 19
   e. 21

8 9)162,000

   a. 1,600
   b. 1,800
   c. 16,000
   d. 18,000
   e. 19,000

9 4)640

   a. 16
   b. 20
   c. 110
   d. 160
   e. 210

10 40)640

   a. 11
   b. 16
   c. 110
   d. 120
   e. 160

Solve these problems. Check your answers to see if they make sense.

11 The grocery store sells a package of 8 oranges for $3.76 (376¢), or charges 50¢ for a single orange. Which is the better value? Explain your answer.

_____

12 A half-marathon is about 13 miles. Josh ran a half marathon in about 2 hours. What was his average speed per hour?

_____

Name _____ Date _____

Using a Bar Graph

Northridge School District is raising money for new computer equipment by selling pizzas. This bar graph shows how many pizzas each grade has sold so far.

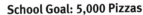

School Goal: 5,000 Pizzas

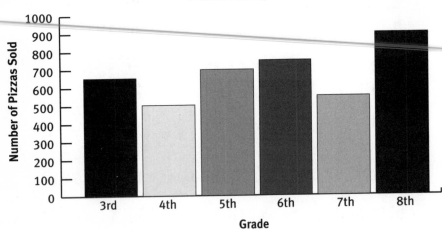

Use the graph to answer these questions.

1 How many pizzas has the 6th grade sold? _____

2 Which grade has sold the most pizzas so far? _____

3 How many pizzas has that grade sold? _____

4 Which grade has sold 500 pizzas? _____

5 How many pizzas have been sold altogether? _____

6 What is the average number of pizzas sold per grade?

_____

7 How many more pizzas must be sold for the school to meet its goal? _____

8 Ms. Westerfeld says that if each grade can sell at least 150 more pizzas, the school will meet its goal. Is she correct? Explain your answer.

_____

_____

Interpreting Circle Graphs

Use the graphs to answer these questions.

Favorite Gym Activities

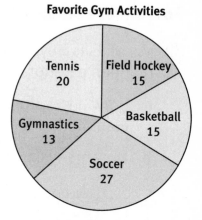

❶ The gym teacher surveyed his students about their favorite gym activity. This circle graph shows the results.

   a. How many students did the gym teacher survey?

   _____

   b. Which activity got the least votes?

   _____

   c. Which activity got about $\frac{1}{3}$ of the votes?

   _____

   d. If we broke the circle graph into 90 equal slices or sectors, what would be the measure, in degrees, of each sector?

   _____

   e. How many degrees are in the angles for each category?

   _____

   _____

❷ The art teacher asked her class of 36 students to choose one of four different projects to work on for the school art contest. The results are shown in the circle graph.

   a. Isabella chose the same project as one-fourth of the class. Which project did Isabella choose?

   _____

   b. Which project did half the students choose?

   _____

   c. How many students chose to do a painting for the art contest? Explain how you found your answer.

   _____

Art Projects

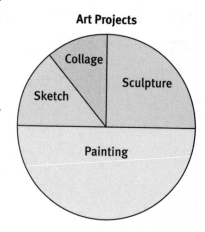

Tree Diagrams

Solve these problems.

❶ Mr. Mills is buying a new car. He can select a black, red, or green exterior color, paired with a tan or gray interior.

   a. Draw a tree diagram to show the different car color combinations.

   b. How many car color combinations are possible?

   _____

   c. Mr. Mills can also select different features for the car. He can pick from the standard, extra, or premium packages. Now, how many different combinations are there?

   _____

❷ At Yogurt Express you can make your own sundae. You can choose between three different flavors of frozen yogurt, paired with hot fudge, peanut butter, strawberry, or caramel topping.

   a. Draw a tree diagram to show all the different kinds of sundaes you can make using one flavor of frozen yogurt and one topping.

   b. If you can also choose between adding peanuts or not adding peanuts, how many different sundaes can you make?

   _____

Graph City

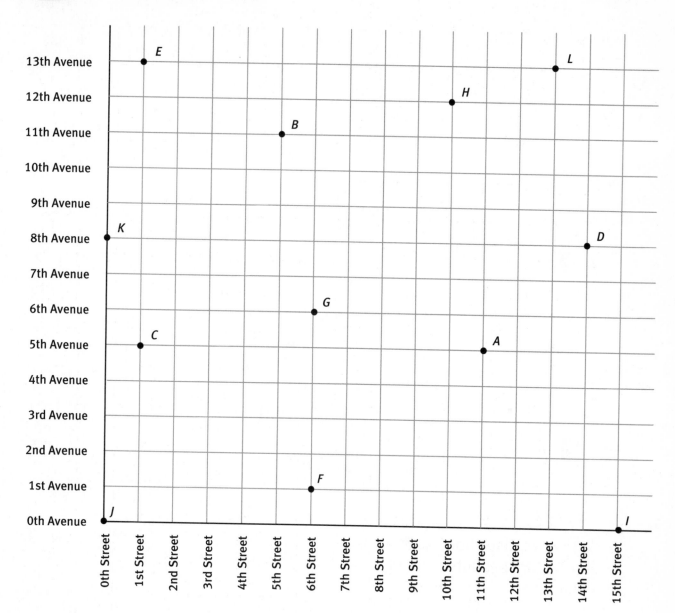

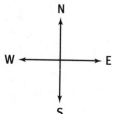

Name _____ **Date** _____

Centimeter Graph Paper

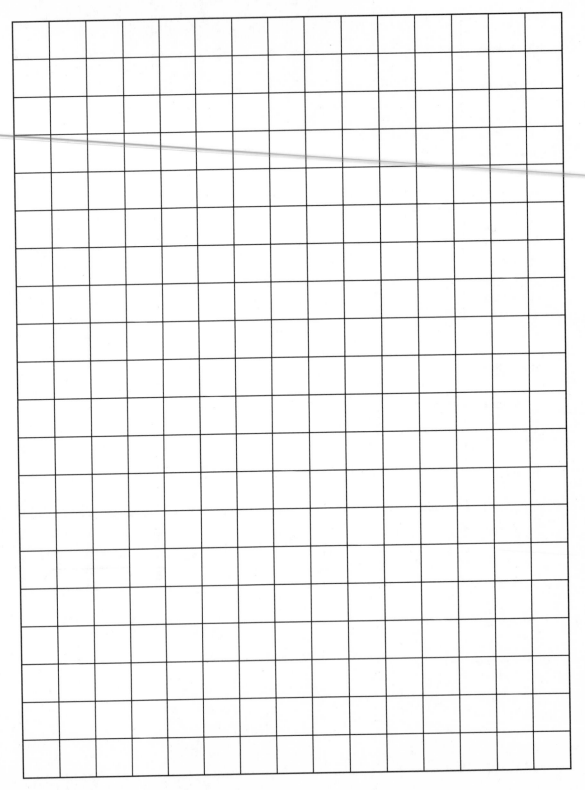

Real Math • Grade 4 • *Practice*

Space Figures

Cut out these polygon patterns to make space figures.

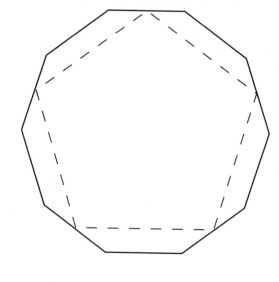
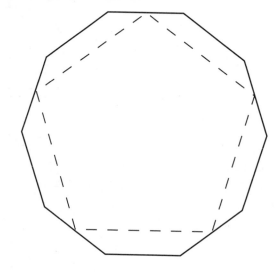

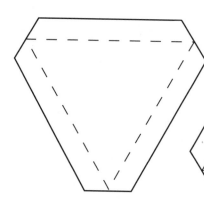
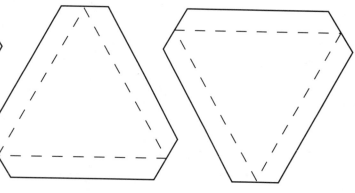

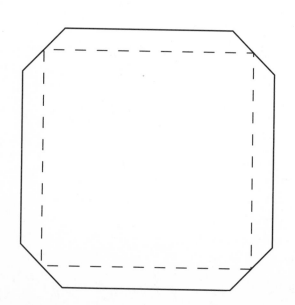
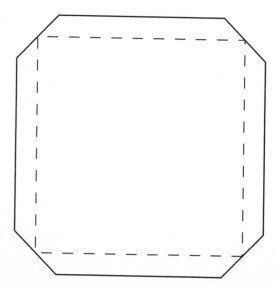

Number Line

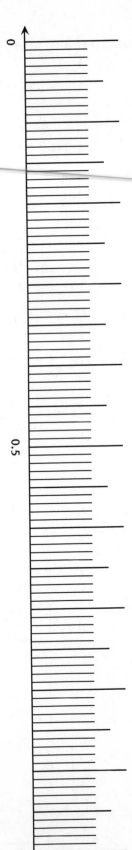

0

0.5

1

Name _____ Date _____

24-Centimeter Strips

Name _____ Date _____

Tens Chart and Hundred Chart

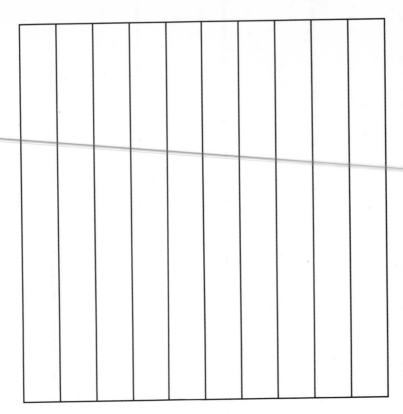

Real Math • Grade 4 • *Practice*